Molecular Theory of Fluctuation in Life Phenomena

Fumio Hirata

Molecular Theory of Fluctuation in Life Phenomena

Springer

Fumio Hirata
Institute for Molecular Science
Okazaki, Japan

ISBN 978-981-96-3363-0 ISBN 978-981-96-3364-7 (eBook)
https://doi.org/10.1007/978-981-96-3364-7

This Springer imprint is published by the registered company Springer Nature Singapore Pte Ltd.
The registered company address is: 152 Beach Road, #21-01/04 Gateway East, Singapore 189721, Singapore

Preface

The word, "FLUCTUATION", in the title of the book may sound somewhat mysterious in the biophysical science. However, the phenomena are physical processes that give *life* to *material*. If there is no fluctuation, molecules in a cell will stay *still* in chemical and/or physical equilibrium *forever*, that means it's *dead*.

The redox reaction occurring in the photosynthesis center in a green plant is a typical example of the process driven by the structural fluctuation of a protein and the density fluctuation of water. The molecular recognition process taking place in the many chemical as well as physical processes in a cell is another example of the process driven by *fluctuation*.

The fluctuation is not a deterministic process, but a probabilistic one that can be described mathematically only by a *probability distribution* of events, such as the conformation of protein and the density of water. There are two types of fluctuation, temporal and spatial. The temporal fluctuation is experimentally measured by photospectroscopy along the frequency as an order parameter, while the spatial fluctuation is measured by diffraction of light (X-ray) or scattering of particles (neutrons) along the wave vector. The two types of fluctuations, one in temporal domain and the other in spatial domain, are not independent each other. Those are related by a law so-called "dispersion relation". For examples, the structural fluctuation of protein has a hierarchical order both in space and time: so-called collective modes extending all over the molecule are characterized by low frequency and small wavenumber, while the oscillation of bond lengths and angles has high frequency and large wavenumber. As another example, the frequency against wave vector concerning the density fluctuation of water gives a linear plot at small wavenumber region, the slope of which gives the *sound velocity*.

This book is not meant as a textbook for students in the physicochemical and biophysical fields, because it does not include sufficient introduction to mathematical and physical background for those fields. It is rather meant for scientists and researchers in the field of the biophysical science, who are seeking good theoretical tools to explore unsolved problem in the field of physics and chemistry. Especially, the book will provide benefits to computational scientists in the field of biophysics, who are trying to set up a good research target for their grant proposal, since many of the theoretical tools proposed in the book are *bland new*, meaning that they have not yet been explored by computational chemistry and physics.

From the view point of the computational science, there are three elements of computational algorithm to be involved: (1) the molecular mechanics and dynamics, (2) the RISM/3D-RISM equation solver, and (3) the eigenvalue problem solver. The first two of those have been well developed in the field of statistical mechanics of liquid and solutions, and enough computational tools have been stocked in the field of computational chemistry, for example, AMBER. On the other hand, the algorithm to solve the eigenvalue problem is well established by the computational scientists in the fields of the quantum physics and chemistry, represented by the Gaussian program originally developed by Pople and his coworkers. It is just a matter of combining those computational algorithms together to solve to the essential problem being unsolved in the life phenomena.

Okazaki, Japan Fumio Hirata

Competing Interests The author has no competing interests to declare that are relevant to the content of this manuscript.

Contents

Chapter 1
Introduction

The unit of life is a cell. A human body contains approximately ~ 10^{14} cells, each of which is composed of chemical species grouped roughly into water, inorganic ions, sugar, amino acids, nucleotide, fatty acids, and other small molecules, macromolecules including protein and nucleic acid. The composition of the chemical species in each cell is different from organ to organ in a living body, for examples, lung, eye, skin, and so on. Chemical composition in a bacterial cell is listed in Table 1.1 as an example (Alberts et al. 1989). Each chemical species plays its own role to maintain our life. DNA carries genetic information to be inherited from generation to generation. Protein plays many different roles to maintain the life: catalyzing chemical reactions including ATP hydrolysis, signal transduction among cells, and so on.

From a theoretical view point to describe molecular processes taking place in a cell, there is a crucial requirement to be accounted for. That is *inhomogeneity* or *non-uniformity* in space as well as in time, which rejects any phenomenological description such as the continuum models for the thermodynamic equilibrium and the fluid dynamics. If one tries to describe the spatial and/or temporal fluctuations of aqueous solution around a biomolecule with such models, they require boundary conditions as many as the number of atoms included in a biomolecule, since the environment of solvent or fluid around each atom is entirely different from each other (Hirata 2020).

For the equilibrium properties, or the solvation free energy, we already have strong theoretical tools based on the integral equation theory of liquid, represented by the RISM/3D-RISM theory, which can account for the inhomogeneity of the field around a biomolecule (Hirata 2020). Let us show, as an example, the theoretical representation of equilibrium distribution of fluids in inhomogeneous environment of a biomolecule, taking a potassium channel (KcsA), one of protein families called "molecular transporters". The potassium channel has a general architecture consisting of several alpha-helices which penetrate the cell membrane to make a pore at the center of the helices bundle, through which small molecules such as water,

F. Hirata, *Molecular Theory of Fluctuation in Life Phenomena*,
https://doi.org/10.1007/978-981-96-3364-7_1

Table 1.1 Approximate chemical composition of a bacterial cell

	Percentage of total cell weight	Number of types of each molecule
Water	70	1
Inorganic ions	1	20
Sugar and precursors	1	250
Amino acids and precursors	0.4	100
Nucleotides and precursors	0.4	100
Fatty acids and precursors	1	50
Other small molecules	0.2	~ 300
Macromolecules (proteins, nucleic acids)	26	~ 3000

ions, and some organic molecules can be transported through the membrane in and out (Hille 2001).

Depicted in Fig. 1.1a is the equilibrium distribution of water (O, H) and alkali ions (Li^+, Na^+, K^+) in the selective filter region of the ion channel, calculated by the RISM/3D-RISM theory (Phongphanphanee et al. 2014a, b). The series of spots along the channel axis colored with blue to red represent the density distributions of ions and water molecules around the equilibrium positions. The figures indicate clearly that the field inside the channel is extremely inhomogeneous in atomic scale. Such inhomogeneity of the field can be seen more clearly in the free energy profile of ions, projected along the central axis of the channel, which is depicted in Fig. 1.1b for Na^+ and K^+ ions. The free energy profiles have an alternate series of basins and tops along the central axis. The free energy profile suggests that in the equilibrium without any external field, the ions are just fluctuating inside a basin. When an external field is turned on between inside and outside of the membrane, the ions get a bias force to overcome the mountain top to produce an electric current. So, it will be understood that the depth of the basin or the height of the top is important factor to determine the conductivity of ions through the channel, which is the biological function of this protein.

We have just emphasized the importance of the *density fluctuation* of the ions and water inside the channel protein for the transmembrane permeation of ions. However, it is not the whole story. There is another important *fluctuation* to be considered to complete the description, that is, the *structural fluctuation* of the protein itself; KcsA in this particular case. Actually, the KcsA channel protein is buried in a cell membrane, which will give a significant influence on the structure and fluctuation of the protein. So, the structural fluctuation of the protein-membrane complex should be considered in order for complete descriptions of the ion conductivity. Then, let us guess possible effects of the structural fluctuations onto the free energy profile of ions along the central axis of pore. The effects will make the profile a little smoother, that is, the height of the top becomes steeper, while the depth of the basin shallower. As a result, the structural fluctuations will work toward the direction of increasing

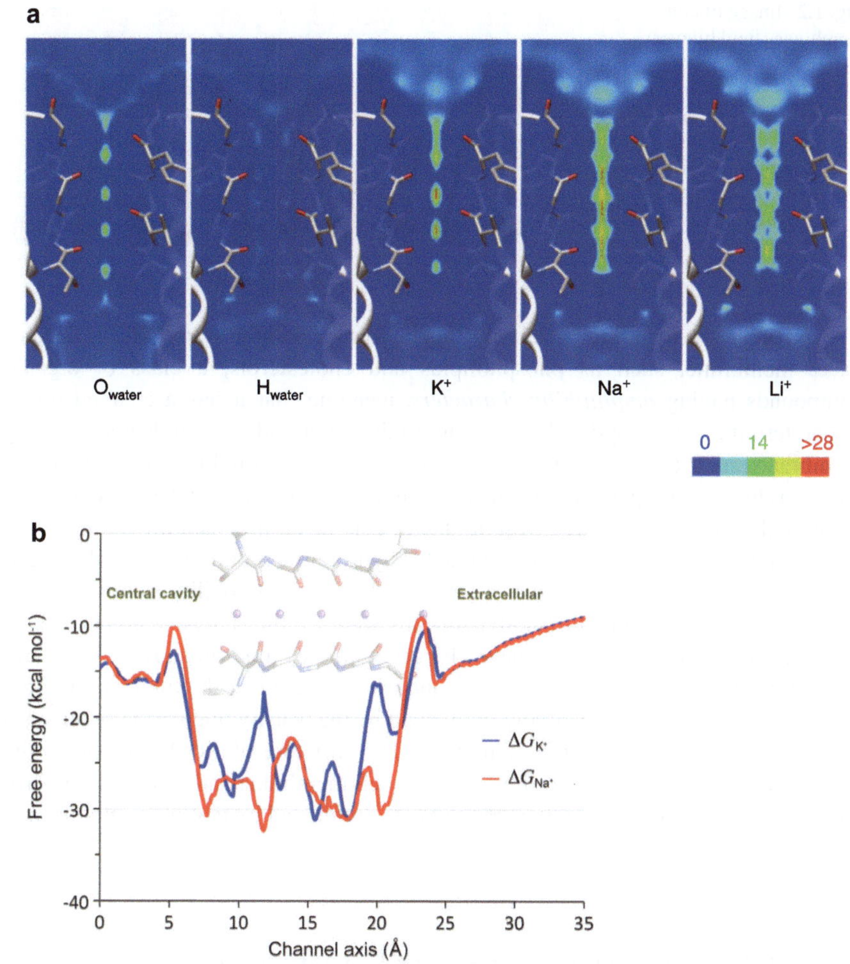

Fig. 1.1 **a** Distribution of water and alkali ions inside the selective filter region of a potassium channel calculated by the RISM/3D-RISM theory (intensity of the distribution is distinguished by color). **b** Potential of mean force (free energy) of K^+ and Na^+ along the central axis of channel pore: red, Na^+; blue, K^+

the ion conductivity through the membrane. There is another important factor that controls the ion conductivity called "gating", which is also a structural fluctuation of the protein (Hille 2001).

We so far considered the effect of structural fluctuation of the protein onto the density fluctuation of solvent, or water and ions. Then, how about the opposite direction? How are the equilibrium structure of a protein-membrane complex and its fluctuation affected by the solvent?

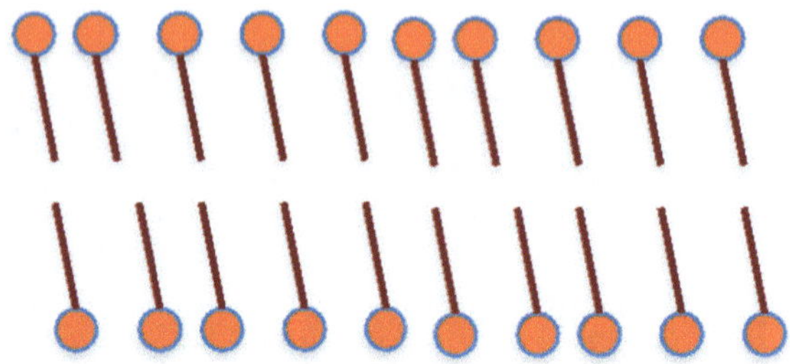

Fig. 1.2 Image of cell membrane (lipid bilayer)

First of all, the membrane itself may not be formed without water. The molecules make membrane, such as phospholipids and cholesterols, a class of organic compounds having *amphiphilic characters*, meaning that it has a charged group in one terminus and a hydrophobic group in the other end. The molecules making membrane, such as phospholipids and cholesterols, are amphiphilic compounds, having a charged group in one terminus and a hydrophobic group in the other end. The amphiphilic molecules make a vesicle-like double layer, in which the hydrophobic groups align tail-to-tail in the layer, while the hydrophilic groups align head-to-head outside the layer (Fig. 1.2). Such a structure having hydrophilic groups aligning head-to-head will never be thermodynamically stable, if they are isolated in *vacuum*, because the hydrophilic groups have charges with the same (negative) sign. Only *water* can make such a structure stable, and it is for sure that the fluctuation of the membrane structure will be largely dominated by the density fluctuation of aqueous solution in- and outside of the membrane. The structure of KcsA protein, or the alpha-helix bundle, is dominated by both the cell membrane and water. So, it is for sure that the structural *fluctuation* of protein is dominated by that of membrane and water as well.

1.1 Molecular Recognition and Self-organization

We have so far looked at the role of density *fluctuation* of solvent as well as structural *fluctuation* of protein to perform their function by taking the molecular transporter as a specific example. In general, there are two molecular processes that are essential for living bodies to maintain their life, in which the *fluctuation* plays crucial roles. Those are the molecular recognition and self-organization (or relaxation) processes (Lehn 1990; Yoshida et al. 2009; Hirata 2020; Sugita et al. 2021).

The molecular recognition is the process in which a biomolecule such as protein binds a ligand, or a small molecule, at its active site with non-covalent bonds. Ions and water molecules bound at the pore of a potassium channel (KcsA) described in the previous section are examples of the molecular recognition.

More typical example of the molecular recognition is seen in an enzymatic reaction. An enzymatic reaction proceeds essentially in three steps: (1) binding substrate (reactant) molecules at the active site, (2) chemical reaction involving the recombination of atoms, and (3) releasing product molecules from the enzyme to be ready for the next catalytic cycle. The first and third steps in the process are nothing but the molecular recognition and reverse processes, respectively, while the second process is an *ordinary* chemical reaction driven by changes of the electronic structure in reactants. A characteristic of an enzymatic reaction, which distinguishes it from ordinary chemical reactions in solution, is the formation of an enzyme–substrate complex referred to as "Michaelis–Menten complex" (Michaelis and Menten 1913; Yoshida et al. 2009).

Assuming that the rate of complex formation is so quick to reach an equilibrium, Michaelis and Menten have derived the following equation for enzymatic reaction:

$$v = \frac{V_{\mathrm{max}}[S]}{K_{\mathrm{m}} + [S]}, \tag{1.1}$$

where $[S]$, K_{m}, and V_{max} denote, respectively, the concentration of the substrate molecules, the equilibrium constant of the complex formation, and the reaction rate at the initial concentration of substrate. It is the equilibrium constant of the complex formation, K_{m}, that features the Michaelis–Menten theory. The formation of Michaelis–Menten complex is essentially a thermodynamic process governed by the free energy change between the bound and unbound states of a pair of protein and ligand molecules. Inclusion of the equilibrium constant of the complex formation in the rate constant of the chemical reaction features an enzymatic reaction, distinguished from the ordinary chemical reactions in solution.

Although we have picked up just one process, or an enzymatic reaction, as an example of the molecular recognition processes, all other functions that are performed by protein, such as transportation of water molecules and ions through cell membranes, are related in one way or the other to the molecular recognition process (Phongphanphanee et al. 2007, 2008, 2010a, b, 2014a, b). As such, binding of drug molecules to a target molecule is nothing but a molecular recognition process (Imai et al. 2009a, b; Kiyota et al. 2011; Phanich et al. 2016; Sugita and Hirata 2016; Hasegawa et al. 2017; Hayashino et al. 2018; Sugita et al. 2020).

The molecular recognition process in a biological system is distinguished from that in gas phase in one essential point, that concerns "water" or "solvation". In usual situation, one or few water molecules and few small ions are bound or *recognized* at the active site of an enzyme or protein before a substrate molecule is bound. Substrate molecules are most likely solvated by water as well. So, in order for protein to bind substrate molecules, some of the water molecules should be removed or *desolvated* from the active site. The free energy change associated with the *desolvation* process, referred to as "desolvation free energy", constitutes a very important part of the free energy change associated with the formation of a Michaelis–Menten complex (Michaelis and Menten 1913).

The self-organization is another important characteristic of biomolecules. Two prototypical examples of the process are (1) the formation of cell membrane and (2) the protein folding (Anfinsen 1973). The process prepares a construction of molecules in which biomolecules play their functions as an enzyme, a signal transducer, an inhibitor, and so on.

Cell membranes are a sort of lipid bilayers, constructed from several ingredients, including phospholipid as a major component. The phospholipid has an amphiphilic characteristic, in which the hydrophilic head groups consist of phosphate groups, while the hydrophobic tail groups are alkyl chains. Those molecules align to make a membrane in such a way as head-to-head and tail-to-tail close together. As been already noted with Fig. 1.2, such a configuration among lipid molecules may not be stable thermodynamically as a membrane due to the two physical causes: the electrostatic repulsion among head groups and the reduced entropy originated from the ordering of the molecules. Then, why the lipid molecules can self-assemble by themselves? The short answer to the question is *water* or *solvation.*

Another example of the self-organization is the so-called "protein folding". Protein changes its conformation from native state to unfolded state when the thermodynamic condition is changed. However, it recovers its unique native conformation reversibly when the thermodynamic state is returned back to the native condition. The phenomenon called "Protein Folding" is first found by Anfinsen (1973). The process is also counter-intuitive, considering that the molecule should climb up the huge entropy barrier to reach a unique native conformation. Of course, there are interactions among the atoms in protein, either attractive or repulsive, and they may contribute to overcome the entropy barrier. However, it can be readily proved by performing the molecular dynamics (MD) simulation of a protein in vacuum that it is not the case. If one performs the MD simulation of a protein in *vacuum* starting from the structure picked up from the protein data bank (PDB), it will quickly collapse into an unidentified non-native conformation within few picoseconds. But then, why does a protein fold into a native conformation when the thermodynamic condition is brought into the native state? The quick answer to the question is again "water" or "solvation" (Anfinsen 1973; Hirata 2020). The solvation free energy including the electrostatic as well as the hydrophobic interactions involving water molecules plays crucial roles for protein to fold into its native state.

There is another physicochemical process that is concerned with both the molecular recognition and self-organization, which is the *structural fluctuation* (Lehn 1990; Hirata 2020; Yoshida et al. 2009). Under normal condition in living cells, the structure of an enzyme is in spatial as well as temporal fluctuation around its native conformation due to thermal motion. A substrate molecule may not be bound at the active site of many of conformations in fluctuated state, since it may not be comfortable geometrically as well as energetically in terms of the free energy including that of solvation. Two popular models to describe the molecular recognition process taking the structural fluctuation into account are "induced fitting" and "conformation selection" (Johnson 2008; Hammes et al. 2009). "Induced fitting" sees the binding process as a temporal process in which a host molecule opens its *gate* or *mouth* consisting of amino-acid residues for accommodating a guest molecule, induced by a perturbation due to the guest molecule. The *conformational selection*, on the other hand, interprets the binding process in terms of an ensemble of the host molecules, including those having the conformation ready for binding the guest molecule. The binding process is a stochastic process depending on the probability of a guest molecule to find the host molecule in the conformation that is *favorite* for binding.

The structural fluctuation also plays a crucial role in self-organization processes in enzymatic reactions. An enzyme as a catalyst should restore its original structure to complete its reaction and be ready for the next reaction cycle. The process is nothing but a self-organization process. Without this process, the protein would stay in a structure that does not function as an enzyme, and it loses its activity as a catalyst.

1.2 Theoretical Description of *Fluctuation*

In the preceding section, we have looked at the two physicochemical processes, *molecular recognition and self-organization*, which play crucial roles in living bodies to maintain their life. The author has also emphasized the role of *fluctuations* in the two physicochemical processes. Then, one may ask "Is it possible to describe theoretically such a process that is apparently a highly random process?" "Is there any law which governs such a random process?" A quick answer to the question is "yes", and it is the main objective of this monograph to construct such a theory to describe fluctuations in life phenomena from the first principle in physics. Here, let us briefly outline the theory in order to help readers having some idea concerning how random processes acquire a simple law. The thermal equilibrium is defined by the minimum point of the free energy surface, while the fluctuation is defined by the deviation from the thermal equilibrium.

Let us define the *local* density of solvent atoms (water and ions) in aqueous solution and its thermal average, respectively, by $\rho_a(\mathbf{r}, t)$ and $\langle \rho_a \rangle$, where a denotes the specific atom a in a solvent molecule. The density fluctuation can be defined by

$$\delta\rho_a(\mathbf{r}, t) = \rho_a(\mathbf{r}, t) - \langle \rho_a \rangle. \tag{1.2}$$

The structure of protein and its statistical average are denoted by $\mathbf{R}_\alpha(t)$ and $\langle \mathbf{R}_\alpha \rangle$, where $\mathbf{R}_\alpha$ represents the Cartesian coordinate of atom α of the protein. The structural fluctuation of the protein is defined by

$$\Delta\mathbf{R}_\alpha(t) = \mathbf{R}_\alpha(t) - \langle \mathbf{R}_\alpha \rangle. \tag{1.3}$$

It should be noted that the fluctuations described by the two variables are not independent events, but they interplay each other: that is, the structure of protein is largely influenced by solvent density fluctuations, while the density fluctuations of solvent atoms (H, O and ions) are governed by the external field exerted from the protein.

In 2013, Kim and Hirata have derived a couple of generalized Langevin equations which describe the time evolution of $\delta\rho_a(\mathbf{r}, t)$ and $\Delta\mathbf{R}_\alpha(t)$ in the phase space (Kim and Hirata 2013). If the random and frictional forces in the two equations are dismissed, the equations reduce to the following equations which describe the *steady-state* fluctuations of solvent density and the protein structure.

$$\frac{\partial^2 \delta\rho_a(\mathbf{r}, t)}{\partial t^2} = -\sum_{b,c} \int \kappa_{ab}^{\mathrm{solv}}(\mathbf{r}, \mathbf{r}')\delta\rho_c(\mathbf{r}', t)\mathrm{d}\mathbf{r}' \tag{1.4}$$

and

$$M_\alpha \frac{\mathrm{d}^2 \Delta\mathbf{R}_\alpha(t)}{\mathrm{d}t^2} = -\sum_\beta K_{\alpha\beta}\Delta\mathbf{R}_\beta(t). \tag{1.5}$$

The right-hand side in both equations represents the force to restore the equilibrium from the fluctuated state. It may be rather surprising that the restoring force is proportional to the deviation from an equilibrium, or the *fluctuation*, since it means that they are formally equivalent to the equation to describe a *harmonic oscillator*. This finding is extremely important for applying the equations to biomolecules in solutions, because it enables us to carry out *harmonic analyses*, similar to the normal mode analysis (NMA), to a biomolecule in water. The equation also indicates that the probability distribution of the fluctuation is *Gaussian* and that the second moment of the Gaussian distribution, or the variance–covariance matrix, represents the magnitude of the fluctuation (Fig. 1.5).

1.3 Experimental Evidence for the Gaussian Fluctuation of Protein

An experimental evidence for the structural fluctuation to be *Gaussian* has been provided by Kataoka et al. in the paper published in 1995 (Kataoka et al. 1995). The authors have carried out small angle X-ray scattering (SACS) measurement for myoglobin in water in a variety of thermodynamic conditions, *native*, *denatured*, and *molten globule states*, and have plotted the logarithm of the scattering intensity against the square of wave vector, which gives so-called Guinier plot (Fig. 1.3).

The idea of Guinier plot is based on the following equation for the probability distribution of atoms in a molecule, expressed in the Fourier space:

$$P(Q) \propto \exp\left[-kQ^2\right], \tag{1.6}$$

where k is inverse of the *correlation length* of the structural fluctuation, or the displacement of atoms from equilibrium position, $k = 1/\langle \Delta R^2 \rangle$. All the plots corresponding to the different states of protein have exhibited linear behavior with negative slopes in small Q-region: the larger the degrees of denaturation, the greater the slope. The behavior is an unequivocal evidence for the distribution of the structural fluctuation in smaller Q-region being "*Gaussian*". Since the small Q-region of structural

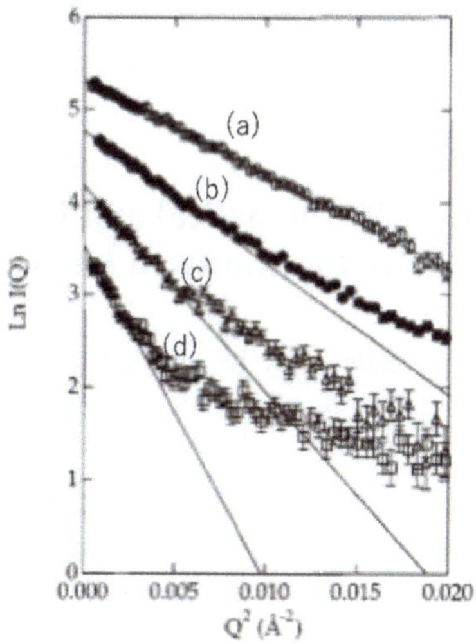

Fig. 1.3 Guinier plot for a variety of conformations of myoglobin:
a homomyoglobin (native state), **b** apomyoglobin (native state), **c** molten globule state, **d** unfolded state

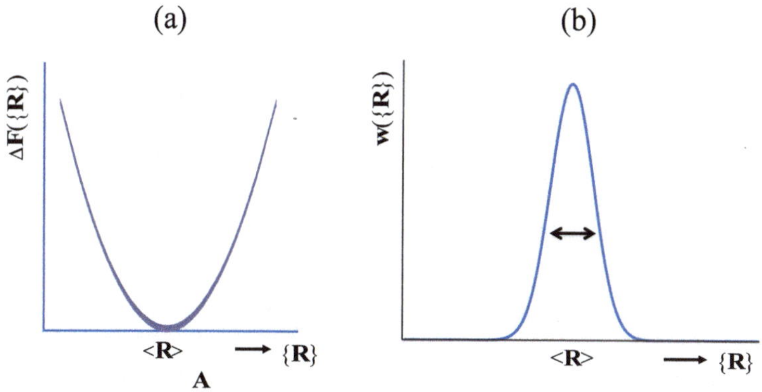

Fig. 1.4 **a** Schematic view of the free energy surface $\Delta F(\{\mathbf{R}\})$ of the structural fluctuation, which is a quadratic surface. The curvature of surface $K_{\alpha\beta}$ is defined by Eq. (1.14). **b** The probability distribution of the structural fluctuation $w(\{\mathbf{R}\})$ which is Gaussian. The half-width of the distribution, or the variance–covariance matrix, is defined by Eq. (1.16)

fluctuations corresponds to a *collective* motion, the results indicate that the collective fluctuation of protein is "Gaussian" irrespective of the states of protein, *native*, *molten globule*, or *denatured*. The collective fluctuation of native conformation has the least k, and the same slope extends over the wide range of Q. It is because the structural fluctuation of the native state, represented by $\langle \Delta R^2 \rangle$, is smallest among all the conformational states of protein. There is another interesting feature seen in the plot of the denatured states of protein, that is the change in k around $Q \sim 0.005$ Å^{-2}. Such larger Q corresponds to rather local region of the structure. Therefore, the change in k reflects the transition in the mode from the collective motion to a rather local mode such as oscillatory motion of individual amino-acid residues.

1.4 Mathematical and Physical Background of the Gaussian Fluctuation

The reader may be puzzled how and why the structural fluctuation of protein in water can be a Gaussian process. There are two reasons to answer the question: one mathematical and the other physical.

Central Limiting Theorem: The mathematical answer to the question is the *central limiting theorem*, a theorem in statistics which can be proved mathematically for any phenomena having large number of fluctuating variables or events (Chandrasekhar 1943; Kubo et al. 1991).

The theorem says that randomly fluctuating variables takes the Gaussian distribution, or the normal distribution, around its average value, unless some of the fluctuations is extraordinarily large. The theorem is rigorously proved at the limit of infinitely large number of variables. In the case of a protein in *vacuum*, the number of variables (atoms) is not so large, ~ 10^4, for the central limiting theorem to be satisfied. On the other hand, a protein in water has essentially an infinite number of degrees freedom, due to that of water molecules, the number of which is ~ 10^{23}/mol. So, the structural fluctuation of protein in water is dominated by the degrees of freedom of solvent, which is overwhelmingly large.

Fluctuation of Thermodynamic Variables: The physical answer to the question is concerned with the fluctuation of thermodynamic variables (Landau and Lifshitz 1958).

Thermodynamic variables, energy, free energy, temperature, density, and so on are fluctuating around their average values due to the thermal fluctuation. Such a situation can be expressed by the Taylor expansion of the free energy of a system around its equilibrium value as

$$F(T,V) = F(T_0,V_0) + \left(\frac{\partial F}{\partial T}\right)_V \Delta T + \left(\frac{\partial F}{\partial V}\right)_V \Delta V + \frac{1}{2}\left(\frac{\partial^2 F}{\partial T^2}\right)_V \Delta T^2 + \frac{1}{2}\left(\frac{\partial^2 F}{\partial T \partial V}\right)\Delta T \Delta V + \frac{1}{2}\left(\frac{\partial^2 F}{\partial V^2}\right)_T \Delta V^2 + \cdots,$$

where $F(T_0, V_0)$ is the free energy of the system at equilibrium, and $\Delta T (\equiv T - T_0)$ and $\Delta V (\equiv V - V_0)$ denote the fluctuation of temperature and volume of the system, respectively. The first derivative of the free energy should be zero by definition of the equilibrium. Therefore, the fluctuation of free energy is expressed as,

$$\Delta F \equiv F - F_0 = \frac{1}{2}\left(\frac{\partial^2 F}{\partial T^2}\right)_V \Delta T^2 + \frac{1}{2}\left(\frac{\partial^2 F}{\partial T \partial V}\right)\Delta T \Delta V + \frac{1}{2}\left(\frac{\partial^2 F}{\partial V^2}\right)_T \Delta V^2. \tag{1.7}$$

Applying the standard thermodynamic manipulation to the derivatives, one finds,

$$\Delta F = -\frac{1}{2}\frac{C_V}{T}\Delta T^2 - \frac{1}{2}\beta \Delta T \Delta V + \frac{1}{V\chi_T}\Delta V^2, \tag{1.8}$$

where C_V, β, χ_T are the heat capacity, the thermal coefficient, and the isothermal compressibility defined by

$$C_V = T\left(\frac{\partial S}{\partial T}\right), \quad \beta = \left(\frac{\partial V}{\partial p}\right), \quad \chi_T = -\frac{1}{V}\left(\frac{\partial V}{\partial p}\right)_V. \tag{1.9}$$

The probability distribution of the fluctuation of free energy may be expressed as

$$\begin{aligned} w(\Delta F) &= C \exp\left[-\frac{\Delta F}{k_B T}\right] \\ &= C \exp\left[\frac{C_V}{2k_B T^2}\Delta T^2 + \frac{\beta}{2k_B T}\Delta T \Delta V - \frac{1}{2Vk_B T \chi_T}\right]. \end{aligned} \tag{1.10}$$

Applying the theorems regarding the Gaussian distribution, one finds the following relation between the thermodynamic variables and the variance–covariance matrix of the fluctuation:

$$\langle \Delta T^2 \rangle = \frac{k_B T^2}{C_V}, \quad \langle \Delta T \Delta V \rangle = \frac{k_B T}{\beta}, \quad \langle \Delta V^2 \rangle = V k_B T \chi_T. \tag{1.11}$$

Structural Fluctuation of a Biomolecule in Water: The conformational stability of a biomolecule in aqueous solution is determined not just by the interaction energy among atoms in the biomolecule, but by the entire free energy including both the interaction energy and the solvation free energy.

Let us denote the free energy, the interaction energy, and the solvation free energy, respectively, by $F(\{\mathbf{R}\})$, $U(\{\mathbf{R}\})$, and $\Delta\mu(\{\mathbf{R}\})$, where $\{\mathbf{R}\}$ represents the set of coordinates of atoms, or the *structure* of a biomolecule. The free energy of a biomolecule in water may be defined by

$$F(\{\mathbf{R}\}) \equiv U(\{\mathbf{R}\}) + \Delta\mu(\{\mathbf{R}\}). \tag{1.12}$$

The free energy is expressed by the Taylor expansion of the fluctuation $\{\Delta\mathbf{R}\}$ around its equilibrium structure as

$$F(\{\mathbf{R}\}) = F(\langle\mathbf{R}\rangle) + \sum_{\alpha} \frac{\partial F}{\partial \mathbf{R}_\alpha} \Delta\mathbf{R}_\alpha + \frac{1}{2}\sum_{\alpha}\sum_{\beta} \frac{\partial^2 F}{\partial \mathbf{R}_\alpha \partial \mathbf{R}_\beta} \Delta\mathbf{R}_\alpha \Delta\mathbf{R}_\beta + \cdots ,$$

where α and β denote the atomic coordinates of the biomolecule, and $\langle\mathbf{R}\rangle$ represents the atomic coordinates of protein in its equilibrium structure.

The first derivative, $(\partial F/\partial \mathbf{R}_\alpha)$, vanishes at equilibrium by definition. The free energy change due to structural fluctuation, $\Delta F(\{\mathbf{R}\}) \equiv F(\{\mathbf{R}\}) - F(\langle\mathbf{R}\rangle)$, is expressed as

$$\Delta F(\{\mathbf{R}\}) = \frac{1}{2}\sum_{\alpha}\sum_{\beta} K_{\alpha\beta} \Delta\mathbf{R}_\alpha \Delta\mathbf{R}_\beta, \tag{1.13}$$

where the force constant, or *Hessian*, is defined by

$$K_{\alpha\beta} = \frac{\partial^2 F}{\partial \mathbf{R}_\alpha \partial \mathbf{R}_\beta}. \tag{1.14}$$

In the natural condition of living system, the higher order terms in the expansion should be ignored, unless otherwise the living system will be *dead*.

Then, the probability distribution of the fluctuation $\Delta\mathbf{R}_\alpha$ may be expressed by

$$\begin{aligned} w(\{\mathbf{R}\}) &= C \exp\left[-\frac{\Delta F(\{\mathbf{R}\})}{k_\mathrm{B} T}\right] \\ &= C \exp\left[-\frac{1}{2k_\mathrm{B} T}\sum_{\alpha}\sum_{\beta} K_{\alpha\beta} \Delta\mathbf{R}_\alpha \Delta\mathbf{R}_\beta\right], \end{aligned} \tag{1.15}$$

where C is the normalization constant.

The equation indicates that the distribution is *Gaussian* in the multi-dimensional space spanned by the atomic coordinates of the biomolecule. The dispersion, or the variance–covariance matrix, of the distribution is identified as

$$\left\langle \Delta\mathbf{R}_\alpha \Delta\mathbf{R}_\beta \right\rangle = \left(\frac{1}{2k_\mathrm{B} T}\frac{\partial^2 F}{\partial \mathbf{R}_\alpha \partial \mathbf{R}_\beta}\right)^{-1}. \tag{1.16}$$

The free energy surface and the Gaussian distribution are schematically illustrated in Fig. 1.4.

It should be noted that the above argument does not apply to the *interaction energy* $U(\{\mathbf{R}\})$, because $\partial U/\partial \mathbf{R}_\alpha$ is not zero at the conformational equilibrium of a biomolecule in aqueous solution. It is the solvent or *water* that makes the conformational fluctuation of a biomolecule *Gaussian*. The conformational fluctuation is dominated by the overwhelmingly large degrees of freedom of solvent, which meets the mathematical condition of the *central limiting theorem*.

1.5 Chemical Equilibrium and Kinetics Described with the Molecular Theory of Fluctuation

If the thermodynamic condition, such as temperature, pressure, and volume, of solvent environment around a biomolecule is changed, not only the equilibrium structure $\langle \mathbf{R} \rangle$ but also the variance–covariance matrix $\langle \Delta \mathbf{R}_\alpha \Delta \mathbf{R}_\beta \rangle$ will be changed accordingly. Such a conformational change of a biomolecule induced by a change in thermodynamic conditions is nothing but a *chemical reaction*, so-called "isomerization reaction" in chemistry. The situation is illustrated by an image drawn in Fig. 1.5.

In the figure, the two parabolas represent the free energy surface of a biomolecule in water in different thermodynamic conditions, for examples, different pressures. The equilibrium constant of the reaction may be defined by

$$K_{\mathrm{eq}} = \exp\left[-\Delta F^0 / k_{\mathrm{B}} T\right], \tag{1.17}$$

where ΔF^0 is the difference in the free energies of the equilibrium conformations before and after the reaction, that is,

$$\Delta F^0 \equiv F\left(\{\mathbf{R}_{\mathrm{eq}}^{(f)}\}\right) - F\left(\{\mathbf{R}_{\mathrm{eq}}^{(i)}\}\right). \tag{1.18}$$

On the other hand, the *rate* of chemical reaction may be described by the *transition-state theory*, originated by Arrhenius in 1884,

$$k = \upsilon \exp\left(-\frac{\Delta F^{(\dagger)}}{kT}\right), \tag{1.19}$$

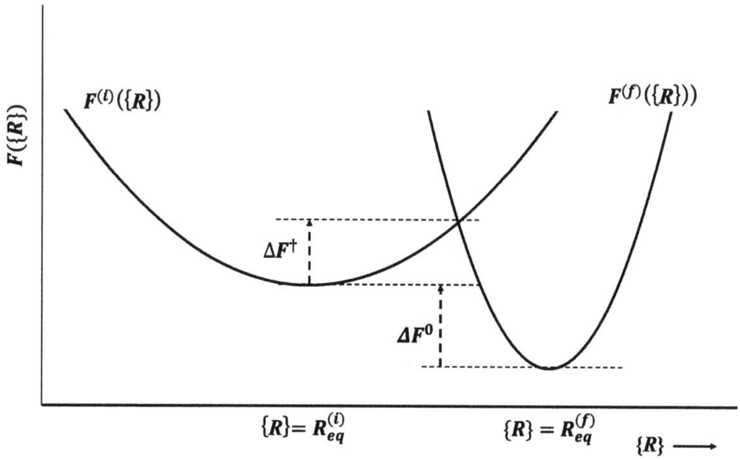

Fig. 1.5 Schematic view of the free energy surface of a biomolecule in chemical reaction

in which $\Delta F^{\dagger}$ is referred to as the *activation free energy* defined by

$$\Delta F^{(\dagger)} \equiv F\left(\{\mathbf{R}^{\dagger}\}\right) - F\left(\{\mathbf{R}_{\mathrm{eq}}^{(i)}\}\right), \tag{1.20}$$

where $\{\mathbf{R}^{\dagger}\}$ symbolizes the coordinates of the *seam* of the two quadratic surfaces in the multi-dimensional space (Arrhenius 1884).

References

Alberts B, Bray D, Lewis J, Baff M, Roberts K, Watson JD (1989) Molecular biology of the cell. Garland Publishing, Inc., New York, London
Anfinsen CB (1973) Principle that govern the folding of protein chain. Science 181:223–230
Arrhenius S (1884) Recherches sur la conductivité galvanique des électrolytes. Doctoral dissertation, Royal Publishing House, Stockholm. P.A. Norstedt & Söner, 89 pp
Chandrasekhar S (1943) Stochastic problems in physics and astronomy. Rev Mod Phys 15:1–89
Hammes GG, Chang Y-C, Oas TG (2009) Conformational selection or induced fit: a flux description of reaction mechanism. PNAS 106:13737–13741
Hasegawa T, Sugita M, Kikuchi T, Hirata F (2017) A systematic analysis of the binding affinity between the Pim-1-kinase and its inhibitors based on the MM/3D-RISM/KH method. J Chem Inf Model 57:2789–2798
Hayashino Y, Masatake S, Hidetoshi A, Irie T, Kikuchi T, Hirata F (2018) Predicting the binding mode of 2-hydroxypropyl-beta-cyclodextrin to cholesterol by means of the MD simulation and the 3D-RISM-KH theory. J Phys Chem B 122:5716–5725
Hille B (2001) Ion channels of excitable membranes, 3rd edn. Sinauer
Hirata F (2020) Exploring life phenomena with statistical mechanics of molecular liquids. CRC Press, New York
Imai T, Kovalenko A, Hirata F, Kidera A (2009a) Molecular thermodynamics of trifluoroethanol-induced helix formation: analysis of the solvation structure and free energy by the 3D-RISM theory. Interdiscip Sci Comput Life Sci 1:156–160
Imai T, Oda K, Kovalenko A, Hirata F, Kidera A (2009b) Ligand mapping on protein surfaces by the 3D-RISM theory; toward computational fragment-based drug design. J Am Chem Soc 131:12430–12440
Johnson K (2008) Role of induced fit in enzyme specificity: a molecular forward/reverse switch. J Biol Chem 283:26297–26301
Kataoka M, Nishi I, Fujisawa T, Ueki T, Tokunaga F, Goto Y (1995) Structural characterization of the molten globule and native states of apomyoglobin by solution X-ray scattering. J Mol Biol 249:215–228
Kim B, Hirata F (2013) Structural fluctuation of protein in water around its native state: a new statistical mechanics formulation. J Chem Phys 138:054108–054111
Kiyota Y, Yoshida N, Hirata F (2011) Affinity of small ligands to myoglobin studied by the 3D-RISM theory. J Mol Liq 159:93–98
Kubo R, Toda M, Hashitsume N (1991) Statistical physics II: nonequilibrium statistical physics. Springer
Landau LD, Lifshitz EM (1958) Statistical physics. Addison-Wesley, Reading, MA
Lehn J-M (1990) From molecular recognition towards molecular information processing and self-organization. Angew Chem 29:1304–1319
Michaelis L, Menten M (1913) Die kinetik der invertinwirkung. Biochem Z 49:333–369
Phanich J, Rungrotmongkol T, Sindhikara D, Phongphanphanee S, Yoshida N, Hirata F, Kungwan N, Hannonbua S (2016) A 3D-RISM/RISM study of the oseltamivir binding efficiency with

the wild-type and resistance-associated mutant forms of the viral influenza B neuraminidase. Protein Sci 25:147–158
Phonphanphanee S, Yoshida N, Hirata F (2007) A statistical mechanics study on equilibrium water distribution in aquaporin. Chem Phys Lett 449:196–201
Phongphanphanee S, Yoshida N, Hirata F (2008) On the proton exclusion of aquaporins: a statistical mechanics study. J Am Chem Soc (Commun) 130:1540–1541
Phongphonphanee S, Yoshida N, Fumio H (2010a) Molecular selectivity in aquaporin channels studied by the 3D-RISM theory. J Phys Chem B 114:7967
Phongphonphanee S, Rungromongkol T, Yoshida N, Hannongbua S, Hirata F (2010b) Proton transport through the influenza A M2 channel: 3D-RISM study. J Am Chem Soc 132:9782–9788
Phongphanphanee S, Yoshida N, Oiki S, Hirata F (2014a) The 'ambivalent' snug-fit sites in the KcsA potassium channel probed by 3D-RISM theory. Pure Appl Chem 86:97–104
Phongphanphanee S, Yoshida N, Oiki S, Hirata F (2014b) Distinct configurations of cations and water in the selective filter of the KcsA potassium channel probed by the 3D-RISM theory. J Mol Liq 200:52–58
Sugita M, Hirata F (2016) Predicting the binding free energy of the inclusion process of 2-hydroxypropyl-b-cyclodextrin and small molecules by means of the MM/3D-RISM method. J Phys Condens Matter 28:384002–384013
Sugita M, Hamano M, Kasahara K, Kikuchi T, Hirata F (2020) New protocol for predicting the ligand-binding and mode based on the 3D-RISM/KH theory. J Chem Theor Comput 16:2864–2876
Sugita M, Onishi I, Irisa M, Yoshida Y, Hirata F (2021) Molecular recognition and self-organization in life phenomena studied by a statistical mechanics of molecular liquids, the RISM/3D-RISM theory. Molecules 26:271–300
Yoshida N, Imai T, Phongphanphanee S, Kovalenko A, Hirata F (2009) Molecular recognition in biomolecules studied by statistical-mechanical integral-equation theory of liquids. J Phys Chem B 113:873–886

Chapter 2
Spatial Density Fluctuations in Water

The chapter is devoted to review the theories of spatial density fluctuation in aqueous solutions. The density fluctuation is characterized by the *pair correlation function*, which is essentially the variance of the probability distribution of density. In Sect. 2.1, we introduce a physical quantity referred to as "density field" which defines the density of liquids at a *local point*. In Sect. 2.2, we define the density pair correlation function by the variance of density fluctuation. Section 2.3 is concerned with the equation referred to as Ornstein–Zernike (OZ) equation which governs the density pair correlation function. Sections 2.4 to 2.6 are devoted to explain applications of the OZ equation to a class of liquids called "simple liquids" in which molecules are just *spherical*. Section 2.7 is concerned with the RISM theory. The RISM theory is an OZ equation generalized to so-called molecular liquids in which a molecule is not spherical but a composite of atoms. However, the theory is not applicable to water in which intermolecular interactions involve the electrostatic interaction. The RISM theory applicable to water, referred to as the XRISM theory, is explained in Sect. 2.8. The theory to explore the solvation structure of a biomolecule in water, referred to as the 3D-RISM theory, and its application to protein is reviewed in Sects. 2.8–2.11.

2.1 Definition of the Local Density of Liquid: Density Field

Any quantity related to molecules in a living cell is subject to thermal motion, that is a randomly varying event. Such a random event is referred to as *fluctuation*. The fluctuation occurring in our body is classical events, except for those related to the electronic structure, indicating that the probability distribution of the fluctuation is continuous. Such fluctuations are characterized by the two physical quantities: *average* and *variance*.

It is the Gaussian distribution that governs fluctuations in a living system most universally.

F. Hirata, *Molecular Theory of Fluctuation in Life Phenomena*,
https://doi.org/10.1007/978-981-96-3364-7_2

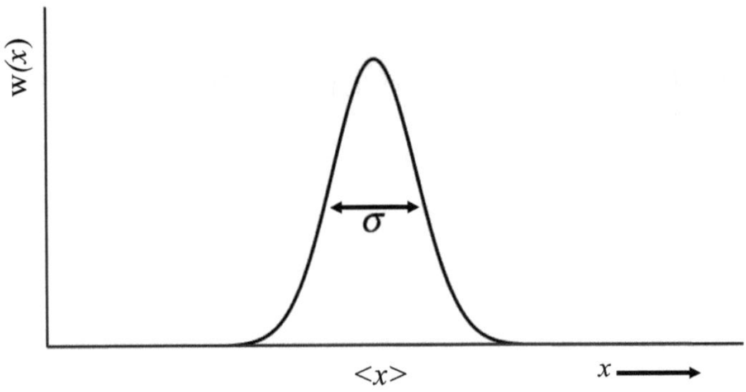

Fig. 2.1 Gaussian distribution in one dimensional space

$$w(x) = \frac{1}{\left(2\pi\sigma^2\right)^{1/2}} \exp\left[-\frac{\Delta x^2}{2\sigma^2}\right] \tag{2.1}$$

In the equation, $\Delta x (\equiv x - \langle x \rangle)$ defines the fluctuation of a random variable x from its average value, or the first moment, $\langle x \rangle$, and σ is the second moment of the distribution, or the variance, defined by $\sigma^2 \equiv \langle \Delta x^2 \rangle$. Illustrated in Fig. 2.1, is a schematic view of the Gaussian distribution in a one-dimensional space.

In the present chapter, we briefly review the theory of water and aqueous solutions from a view point of the spatial density-fluctuation. It is the quantity referred to as "*density pair-correlation function*" that characterize the liquid structure of water. In the following, we construct the theory of the density pair-correlation function from a physical view point of the spatial density fluctuation. In the theory, the density pair correlation function is defined as the variance of the density fluctuation.

A thermodynamic variable that characterizes the liquid state is the density ρ which is defined by the number of molecules, N, divided by the volume V of container, that is, $\rho \equiv N/V$. Then, let us ask a question. What is the density at a *local point* $\mathbf{r}$ in space of liquid system? It is a highly nontrivial question, because the *volume* of a *local point* in space is *zero*. A quick answer to the question is the *Dirac delta function.* Let's find it out first in the one-dimensional space (Harashima 1954; Hirata 2020).

In Fig. 2.2, illustrated by a cartoon is a molecule fixed at position x_i in one dimensional space along the x-axis. Let's denote the density of a molecule at position x by $\nu(x)$, and take an integral of $\nu(x)$ over a range L defined along the x-axis. Then, the integral will be unity if the range L includes the position x_i, otherwise it is zero, that is,

$$\int_L \nu(x)\mathrm{d}x = 1, \; x_i \in L \\ = 0, \; x_i \notin L \tag{2.2}$$

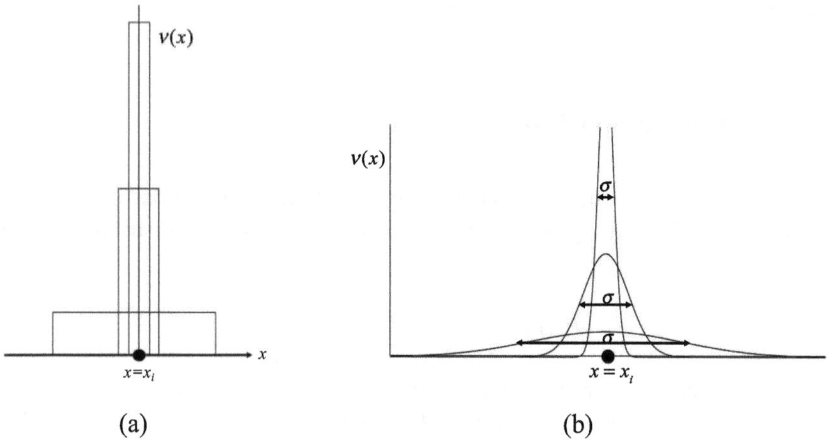

Fig. 2.2 Illustration of the density field in one dimensional space

Now, let the range L be less and less sustaining the condition illustrated in Eq. 2.2, and bring it to the limit $L = 0$. A function which satisfies such a condition is nothing but the *Dirac delta-function*, namely,

$$\nu(x) = \delta(x - x_i) \tag{2.3}$$

The same argument can be made by assuming that the density field $\nu(x)$ around a point x_i in one dimensional space is Gaussian as is illustrated in Fig. 2.2b.

$$\nu(x) = \frac{1}{(2\pi\sigma^2)^{1/2}} \exp\left[-\frac{(x - x_i)^2}{2\sigma^2}\right]$$

Now, take an integral of $\nu(x)$ over the range $-\infty < x < \infty$ along the x-axis. The integral will be unity if the range includes the position x_i,

$$\int_{-\infty}^{\infty} \nu(x)\mathrm{d}x = 1,$$

otherwise the integral is zero. Now, let us reduce the variance σ of $\nu(x)$ less and less to the limit $\sigma \to 0$ with keeping the normalization condition as is illustrated in Fig. 2.2b. The function that satisfies the condition is nothing but the Dirac delta-function.

The definition can be readily generalized to the case where N molecules are fixed in the three-dimensional (3D) space, that is,

$$\nu(\mathbf{r}) = \sum_{i=1}^{N} \delta(\mathbf{r} - \mathbf{r}_i), \tag{2.4}$$

where $\mathbf{r}_i$ ($i = 1, 2, \ldots, N$) define the position of N molecules.

2.2 Spatial Density Fluctuation

According to the statistical mechanics, the probability distribution of mechanical variables in a classical system is governed by the Boltzmann distribution in the *configuration* space,

$$P(\{\mathbf{r}\}) = \exp[-U(\{\mathbf{r}\})/k_\mathrm{B}T]/Z \tag{2.5}$$

where $\{\mathbf{r}\} = (\mathbf{r}_1, \mathbf{r}_2, \ldots, \mathbf{r}_\mathrm{N})$ represents the coordinates of N molecules in the system, $U(\{\mathbf{r}\})$ denotes the potential energy of the system, and the normalization constant, Z, is the partition function in the configurational space defined by,

$$Z = \int_V \int_V \cdots \int_V \exp[-U(\{\mathbf{r}\})/k_\mathrm{B}T]\mathrm{d}\mathbf{r}_1\mathrm{d}\mathbf{r}_2\cdots\mathrm{d}\mathbf{r}_N \tag{2.6}$$

The average or the first moment of density at a point, $\mathbf{r}$, in the system can be defined by,

$$\rho^{(1)}(\mathbf{r}) \equiv \langle\nu(\mathbf{r})\rangle = \int_V \nu(\mathbf{r})P(\{\mathbf{r}\})\mathrm{d}\mathbf{r}_1\mathrm{d}\mathbf{r}_2\cdots\mathrm{d}\mathbf{r}_N. \tag{2.7}$$

From Eqs. (2.4) and (2.7), the average density at position $\mathbf{r}$ is expressed as,

$$\begin{aligned}\rho^{(1)}(\mathbf{r}) = \langle\nu(\mathbf{r})\rangle &= \frac{\int_V\int_V\cdots\int_V \nu(\mathbf{r})\exp[-U(\{\mathbf{r}\})/k_\mathrm{B}T]\mathrm{d}\mathbf{r}_1\mathrm{d}\mathbf{r}_2\cdots\mathrm{d}\mathbf{r}_N}{Z}\\ &= \frac{\int_V\int_V\cdots\int_V \sum_i\delta(\mathbf{r}-\mathbf{r}_i)\exp[-U(\{\mathbf{r}\})/k_\mathrm{B}T]\mathrm{d}\mathbf{r}_1\mathrm{d}\mathbf{r}_2\cdots\mathrm{d}\mathbf{r}_N}{Z}\\ &= \frac{N\int_V\int_V\cdots\int_V \exp[-U(\{\mathbf{r}\})/k_\mathrm{B}T]\mathrm{d}\mathbf{r}_2\cdots\mathrm{d}\mathbf{r}_N}{Z}\end{aligned} \tag{2.8}$$

If the molecules in the system are placed in an inhomogeneous field such as in a *solid crystal*, then average density or the first moment $\rho^{(1)}(\mathbf{r})$ is definitely dependent on the position $\mathbf{r}$ in the crystal, and it will be the primary physical-quantity to define the *structure*. A schematic picture to illustrate such a crystal structure is shown in Fig. 2.3b. The density distribution, $\rho^{(1)}(\mathbf{r})$, is likely to be a Gaussian distribution around the position, $\mathbf{r}$, with a very *narrow* half-width, because a molecule at position $\mathbf{r}$ in the crystal should feel the same average potential energy, the functional form of which is parabola with a large curvature. Such an harmonicity of the potential energy is the basis for the normal mode analysis of crystal structure (McQuarrie 1976).

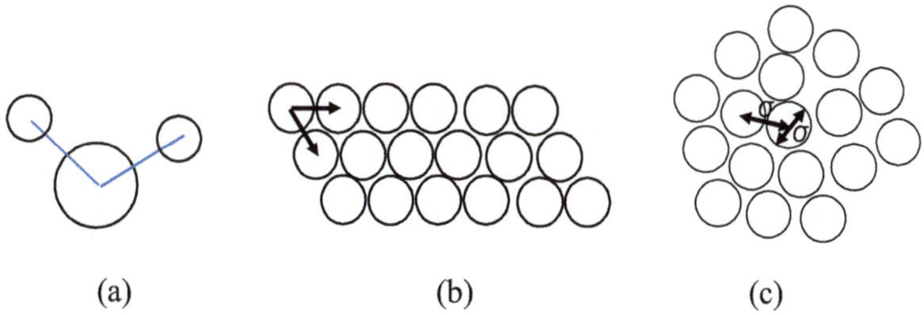

Fig. 2.3 Configuration of molecules (or atoms) in a variety of state of matter: **a** configuration of atoms in a water molecule in gas phase; **b** configuration of molecules in solid state; **c** configuration of molecules in liquid water

On the other hand, the position of each molecule in a *liquid* system is moving around inside the container due to thermal motion. Therefore, one cannot define the position around which a molecule stays for any longer time. The first moment or the average density will be constant.

Considering the translational invariance of the uniform liquid, the coordinate system can be translated to the one in which the origin of the axis is at the center of a molecule, say the molecule 1. The integral concerning the molecule 1 in Eq. (2.8) over the entire space can be readily performed to give,

$$Z = V \int_V \int_V \cdots \int_V \exp[-U(\{\mathbf{r}\})/k_B T] d\mathbf{r}_2 \cdots d\mathbf{r}_N \tag{2.9}$$

Therefore, the average or the first moment of the density filed becomes constant naturally,

$$\langle \nu(\mathbf{r}) \rangle = N/V \equiv \rho \tag{2.10}$$

The result indicates that the average or the first moment of the density filed in liquid is just a constant, and does not carry any *structural* information. Then, how about the second moment, or the variance, of the density field? Let's find it out.

The second moment, or the variance, of the density field is defined by,

$$\rho^{(2)}(\mathbf{r}, \mathbf{r}') \equiv \langle \delta\nu(\mathbf{r})\delta\nu(\mathbf{r}') \rangle, \tag{2.11}$$

where $\delta\nu^{(1)}(\mathbf{r})$ and $\delta\nu^{(1)}(\mathbf{r}')$ are defined by,

$$\delta\nu(\mathbf{r}) \equiv \nu(\mathbf{r}) - \langle \nu(\mathbf{r}) \rangle, \quad \delta\nu(\mathbf{r}') \equiv \nu(\mathbf{r}') - \langle \nu(\mathbf{r}') \rangle \tag{2.12}$$

$$\begin{aligned} \rho^{(2)}(\mathbf{r}, \mathbf{r}') &= \langle \delta\nu^{(1)}(\mathbf{r})\delta\nu^{(1)}(\mathbf{r}') \rangle \\ &= \langle (\nu^{(1)}(\mathbf{r}) - \rho)(\nu^{(1)}(\mathbf{r}') - \rho) \rangle \end{aligned}$$

$$= \langle \nu^{(1)}(\mathbf{r})\nu^{(1)}(\mathbf{r}')\rangle - \rho^2$$

$$= \left\langle \sum_i \delta(\mathbf{r} - \mathbf{r}_i) \sum_j \delta(\mathbf{r}' - \mathbf{r}_j) \right\rangle - \rho^2 \quad (2.13)$$

Let us define a function called the "density pair correlation function" by

$$h(\mathbf{r}, \mathbf{r}') \equiv \rho^{(2)}(\mathbf{r}, \mathbf{r}')/\rho^2. \quad (2.13')$$

The function definitely carries the structural information of liquids, which is observable by experimental means such as the X-ray diffraction and/or the neutron scattering.

The quantity can be also calculated by the statistical-mechanics theory of liquids. The theory is referred to as the "Ornstein–Zernike" equation (Hansen and McDonald 1986).

It is a little involved to derive the equation. Therefore, let us just *construct*, *not derive*, the equation by a physical consideration.

2.3 Ornstein–Zernike Equation

Let us locate the origin of the coordinate at the center of one of molecules, illustrated as in Fig. 2.4, and find the density correlation between a pair of molecules, one of which is fixed at the origin and the other at a position **r** (Fig. 2.4).

Fig. 2.4 Image of configuration of molecules in liquid

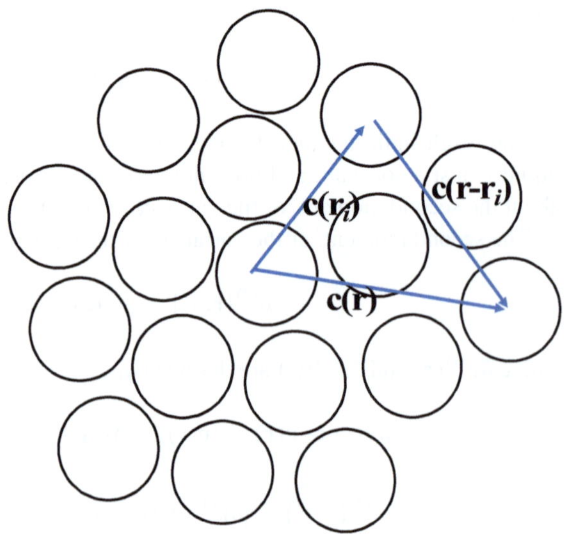

It is worthwhile to note that the *second* moment or variance of the density fluctuation, defined by Eq. (2.3), becomes the *first* moment, or the average density around the molecule fixed at the position $\mathbf{r}'(=0)$ in the space.

The correlation may be conceptually divided into the two contributions, $c(\mathbf{r})$ and $t(\mathbf{r})$,

$$h(\mathbf{r}) = c(\mathbf{r}) + t(\mathbf{r}), \tag{2.14}$$

where $c(\mathbf{r})$ represents the *direct* perturbation of the molecule placed at the origin to that placed at the position $\mathbf{r}$. On the other hand, $t(\mathbf{r})$ reflects the *indirect* perturbation propagated through a molecule placed at the other position $\mathbf{r}_i$: $t(\mathbf{r})$ includes the contributions of $c(\mathbf{r})$ intermediated by molecules at a *series* of positions (see Fig. 2.4)

Let's denote the intermediate positions by $\mathbf{r}_i$ $(i = 1, 2, \ldots, \infty)$. Then, the $t(\mathbf{r})$ may be expressed by an infinite series of $c(\mathbf{r})$,

$$t(\mathbf{r}) = \rho c * c(\mathbf{r}) + \rho^2 c * c * c(\mathbf{r}) + \cdots \tag{2.15}$$

where $c * c(\mathbf{r})$ etc. represents the convolution integral denoted by,

$$c * c(\mathbf{r}) = \int_V c(\mathbf{r})c(\mathbf{r} - \mathbf{r}')\mathrm{d}\mathbf{r}'. \tag{2.16}$$

So, $h(\mathbf{r})$ can be expressed by an infinite series of the direct correlation function as,

$$h(\mathbf{r}) = c(\mathbf{r}) + \rho c * c(\mathbf{r}) + \rho^2 c * c * c(\mathbf{r}) + \cdots \equiv C[c|\rho]. \tag{2.17}$$

In the above equation we have defined $C[c|\rho]$, referred to as "chain sum", by the infinite series of the convolution integrals. The equation can be reorganized mathematically to lead,

$$h(\mathbf{r}) = c(\mathbf{r}) + \rho \int c(\mathbf{r}')h(\mathbf{r} - \mathbf{r}')\mathrm{d}\mathbf{r}'. \tag{2.18}$$

The equation was first derived by Ornstein and Zernike in 1914, and has become a starting point of the modern statistical mechanics of liquids (Ornstein and Zernike 1914).

The expression of the Ornstein and Zernike equation in terms of a series of the convolution integral (Eq. 2.17) has special significance in further development of the liquid state theory, since the convolution integral of two functions becomes merely a product of the Fourier transforms of each function such as,

$$\int \mathrm{d}\mathbf{r} \exp(-i\mathbf{k}\cdot\mathbf{r})\left(\int_V c(\mathbf{r})c(\mathbf{r} - \mathbf{r}')\mathrm{d}\mathbf{r}'\right) = c(\mathbf{k})c(\mathbf{k}). \tag{2.19}$$

Then, the Ornstein–Zernike equation (Eq. 2.17), or the chain sum of $c(\mathbf{r})$, may be written in the Fourier space as,

$$h(\mathbf{k}) = c(\mathbf{k}) + \rho c(\mathbf{k})^2 + \rho^2 c(\mathbf{k})^3 + \cdots \tag{2.20}$$

The sum of the infinite series may be readily performed to give,

$$h(\mathbf{k}) = \frac{c(\mathbf{k})}{1 - \rho c(\mathbf{k})} \tag{2.21}$$

Closure Relation. The Ornstein–Zernike equation includes two variables $h(\mathbf{r})$ and $c(\mathbf{r})$. Therefore, it is necessary to find another equation that relates the two functions, in order for solving the equation. (Such an equation is referred to as "closure relation".) Let us construct such a relation from a physical view point (Hansen and McDonald 1986).

For that purpose, we consider the distribution $\rho(\mathbf{r})$ of molecules at $\mathbf{r}$ in a coordinate system, the origin of which is located at one of molecules (Fig. 2.4). When the density of liquid is very low, like in gas phase, $\rho(\mathbf{r})$ will make the Boltzmann distribution, that is,

$$\rho(\mathbf{r}) \sim \rho \exp[-\,u(\mathbf{r})/k_{\mathrm{B}}T] \tag{2.22}$$

where $u(\mathbf{r})$ denotes the interaction between a pair of molecules.

When the density is higher, the distribution at the position $\mathbf{r}$ will be influenced by the molecules other than those placed at the origin. The effect is separated into two, $t(\mathbf{r})$ and $b(\mathbf{r})$. $t(\mathbf{r})$ is the same function defined by Eq. (2.15) in terms of the convolution integrals, while $b(\mathbf{r})$ is rest of the contribution which cannot be expressed by the convolution integrals. Then, the closure relation can be written as

$$\rho(\mathbf{r}) \equiv \rho g(\mathbf{r}) = \rho \exp[-\,u(\mathbf{r})/k_{\mathrm{B}}T + t(\mathbf{r}) + b(\mathbf{r})], \tag{2.23}$$

where $b(\mathbf{r})$ is conventionally referred as "bridge function". It is the usual approximation in the liquid state theory to ignore the function, because its contribution is small in most of the ordinary thermodynamic conditions of fluids. Then, we get an equation referred as the hyper-netted-chain (HNC) approximation to close the Ornstein–Zernike equation (Eq. 2.18)

$$h(\mathbf{r}) = \exp[-\,u(\mathbf{r})/k_{\mathrm{B}}T + h(\mathbf{r}) - c(\mathbf{r})] - 1 \quad \text{(HNC)}. \tag{2.24}$$

Now, we are ready to find the solution of the two equations, Eqs. (2.18) and (2.24), for $h(\mathbf{r})$ and $c(\mathbf{r})$ of simple liquids, consisting of spherical molecules, given the interaction potential energy between a pair of molecules, $u(\mathbf{r})$.

2.4 Density Fluctuation in Simple Liquids

The potential energy between a pair of atoms in simple liquids can be most popularly represented by the Lennard–Jones interaction, expressed by the following equation,

$$u(r) = 4\varepsilon\left[\left(\frac{\sigma}{r}\right)^{12} - \left(\frac{\sigma}{r}\right)^{6}\right] \tag{2.25}$$

in which σ and ε represent the diameter and the depth of the potential well, respectively.

If you solve the two Eqs. (2.18) and (2.24) together for a liquid system in which the interaction between a pair of molecules is expressed by Eq. (2.25), you may find a curve similar to that shown in Fig. 2.5 for the pair correlation function $g(r)$. The physics of the curve is following.

In the region $r/\sigma < 1$, $g(r)$ is essentially zero due to the repulsive core, the radius of which is σ. On the other hand, $g(r) = 1$, or $\rho g(r) = \rho$, at $r \to \infty$, because the effect of the central atom to the density vanishes, and the liquid density recovers its average value. The first and second peaks at $r/\sigma \simeq 1$ and $r/\sigma \simeq 2$ represent, respectively, the first and second coordination shells around the central molecule (see Fig. 2.4). In such a way, the pair correlation function represents structure of liquids.

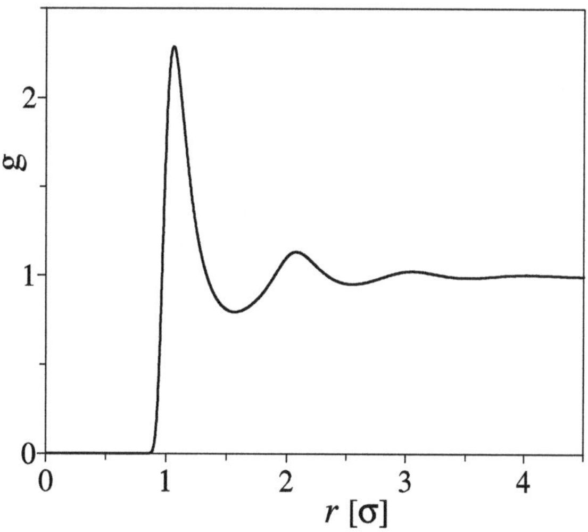

Fig. 2.5 The pair correlation function of a simple liquids interacting with the Lennard–Jones potential. (It should be noted that the x-axis is the radial distance from the origin, reduced by the diameter σ)

2.5 Density Fluctuations in Solutions

It is straight forward to generalize the O–Z equation to a mixture of simple liquids, or solutions. We first express the potential energy with a matrix form by putting suffices that defines component species in the system as,

$$u_{ij}(r) = 4\varepsilon_{ij}\left[\left(\frac{\sigma_{ij}}{r}\right)^{12} - \left(\frac{\sigma_{ij}}{r}\right)^{6}\right] \tag{2.26}$$

where i and j represent the species in the system. The O–Z equation can be expressed accordingly in the matrix form as,

$$h_{ij}(\mathbf{r}) = c_{ij}(\mathbf{r}) + \sum_{l} \rho_l \int c_{il}(\mathbf{r}')h_{lj}(\mathbf{r} - \mathbf{r}')\mathrm{d}\mathbf{r}' \equiv C_{ij}[c|\rho], \tag{2.27}$$

where c_{ij} and h_{ij} denote the direct and pair correlation functions between species i and j, ρ_i represents the density of species i in the system.

The closure relation (Eq. 2.24) is generalized just by writing it in matrix forms as,

$$h_{ij}(\mathbf{r}) = \exp\left[-u_{ij}(\mathbf{r})/k_{\mathrm{B}}T + h_{ij}(\mathbf{r}) - c_{ij}(\mathbf{r})\right] - 1 \tag{2.28}$$

2.6 Density Fluctuations of Ions in Electrolyte Solutions

The simplest model of electrolyte solutions is referred to as "primitive model", in which the interaction between a pair of atoms is expressed as,

$$u_{ij}(r) = u_{ij}^{*}(r) + \frac{z_i z_j}{Dr} \tag{2.29}$$

where $u_{ij}^{*}(r)$ represents the short-range interaction, D is the dielectric constant of solvent represented by a continuum model, and z_i and z_j denote charges on ion species i and j. The charges should satisfy the electro-neutrality condition,

$$\sum_{i} \rho_i z_i = 0. \tag{2.30}$$

It is our problem to solve the O–Z equation (Eq. 2.27) with the HNC approximation (Eq. 2.28) to find the pair correlation function ($h(r)$ or $g(r)$) in Eq. (2.27) for the primitive model of electrolyte solutions (Eq. 2.29). It sounds simple, but in fact it's not, because the integral of the Coulomb interaction over the entire-space of the system diverges, that is,

$$\int_0^{\infty} (1/r)4\pi r^2 \mathrm{d}r \to \infty \tag{2.31}$$

The problem can be solved by changing the order of integration of functions appearing in the chain sum (Eq. 2.17), $C_{ij}[c|\rho]$, so that the chain sum for the Coulomb interaction may be carried out in advance (Rossky and Dale 1980; Friedman 1985). For that purpose, we first split the direct correlation function $c_{ij}(\mathbf{r})$ into two contributions as,

$$c_{ij}(\mathbf{r}) = c_{ij}^{s}(\mathbf{r}) + \phi_{ij}(\mathbf{r}), \tag{2.32}$$

where $\phi(\mathbf{r})$ is the contribution from the Coulomb interaction to $c_{ij}(\mathbf{r})$, defined by,

$$\phi_{ij}(r) = \frac{1}{k_\mathrm{B}T}\frac{z_i z_j}{Dr}, \tag{2.33}$$

and $c_{ij}^{s}(\mathbf{r})$ is the short-range contributions to $c_{ij}(\mathbf{r})$.

The chain sum of ϕ_{ij} can be carried out analytically to give,

$$C_{ij}[\phi|\rho] = -\frac{1}{k_\mathrm{B}T}\frac{z_i z_j}{Dr}\exp[-\kappa r] \equiv q_{ij}(\mathbf{r}) \tag{2.34}$$

where κ is the Debye screening constant defined by

$$\kappa = \frac{4\pi}{Dk_\mathrm{B}T}\sum_i z_i^2 \rho_i. \tag{2.35}$$

The pair correlation function is exactly the same with that obtained by Debye and Huckel from the linearized Poisson–Boltzmann equation for the point-charge model of electrolyte solutions (Debye and Huckel 1923). It should be noted that $q_{ij}(\mathbf{r})$ in Eq. (2.34) is not a long-range interaction unless $\kappa = 0$.

In order to reorganize the chain sum of Eq. (2.17) in terms of $q_{ij}(\mathbf{r})$, we use a function called a *hypervertex function* defined by,

$$\overline{q_{ij}[\phi|\rho]} = \rho_i\delta(\mathbf{r}-\mathbf{r}') + \rho_i\rho_j q_{ij}[\phi|\rho] \tag{2.36}$$

Then, after a little algebra concerning the hypervertex function, one finds,

$$\begin{aligned} h_{ij}(\mathbf{r}) &= C_{ij}[c_\mathrm{s} + \phi|\rho] \\ &= C_{ij}[\phi|\rho] + C_{ij}\left[c_\mathrm{s}|\overline{C[\phi|\rho]}\right] \\ &= q_{ij}(\mathbf{r}) + C_{ij}\left[c_\mathrm{s}|\overline{q}\right] \end{aligned} \tag{2.37}$$

where c_s defines the short-range part of the direct correlation function.

2.7 Density Fluctuation in Molecular Liquids (RISM Theory)

A water molecule is composed of three atoms, oxygen (O) and two hydrogen atoms (H), which are connected by covalent bonds. Let us refer to the atoms in the molecule as "interaction sites", and express the potential energy between two water molecules by a sum of the interaction energies between a pair of water molecules, i and j, by the following expression,

$$u(i,j) = \sum_{a \neq b} u\left(\left|\mathbf{r}_i^a - \mathbf{r}_j^b\right|\right), \tag{2.38}$$

where suffices a and b denote the interaction sites, which are at the oxygen atom (O) and the two hydrogen atoms (H) in case of water. The entire potential energy U of the system is defined by,

$$U = \sum_{i<j} u(i,j) \tag{2.39}$$

The local density of the atom a may be expressed by,

$$\nu_a(\mathbf{r}) = \sum_i \delta(\mathbf{r} - \mathbf{r}_i^a) \tag{2.40}$$

Then, the average density of atom, a, is expressed by the

$$\begin{aligned}
\langle \nu_a(\mathbf{r}) \rangle &= \frac{\int_V \int_V \cdots \int_V \nu_a(\mathbf{r}) \exp\left[-\sum_{i>j} u(i,j)/k_B T\right] d(1)d(2)\cdots d(N)}{Z} \\
&= \frac{\int_V \int_V \cdots \int_V \sum_i \delta(\mathbf{r} - \mathbf{r}_i^a) \exp\left[-\sum_{i>j} u(i,j)/k_B T\right] d(1)d(2)\cdots d(N)}{Z} \\
&= \frac{N_a \int_V \int_V \cdots \int_V \exp\left[-\sum_{i>j} u(i,j)/k_B T\right] d(1)' d(2)\cdots d(N)}{Z}
\end{aligned} \tag{2.41}$$

where $d(i)$ denote the volume element of the integral over the coordinates of atoms in the ith molecule in the system, and $d(1)'$ signifies that the integral with respect to an atom "a" of a molecule "1" is removed from the integral.

As in the case of simple liquids, the density field of atom, a, of a molecule, averaged over the space is constant, and it coincides with the density of molecules.

$$\langle \nu_a(\mathbf{r}) \rangle = N_a/V \equiv \rho_a = \rho \tag{2.42}$$

where ρ denotes the density of molecules in liquids.

The variance of the density distribution is calculated accordingly as,

$$
\begin{aligned}
\rho_{ab}^{(2)}(\mathbf{r},\mathbf{r}') &\equiv \langle \delta\nu_a(\mathbf{r})\delta\nu_b(\mathbf{r}')\rangle \\
&= \langle (\nu_a(\mathbf{r})-\rho)(\nu_b(\mathbf{r}')-\rho)\rangle \\
&= \langle \nu_a(\mathbf{r})\nu_b(\mathbf{r}')\rangle - \rho\langle \nu_b(\mathbf{r}')\rangle - \rho\langle \nu_a(\mathbf{r})\rangle + \rho^2 \\
&= \left\langle \sum_{i=1}^{N}\delta(\mathbf{r}-\mathbf{r}_i^a)\sum_{j=1}^{N}\delta(\mathbf{r}'-\mathbf{r}_j^b)\right\rangle - 2\rho^2 + \rho^2 \\
&= \left\langle \sum_{i=1}^{N}\delta(\mathbf{r}-\mathbf{r}_i^a)\sum_{j=1}^{N}\delta(\mathbf{r}'-\mathbf{r}_j^b)\right\rangle - \rho^2 \\
&= \left\langle \sum_{i=1}^{N}\delta(\mathbf{r}-\mathbf{r}_i^a)\delta(\mathbf{r}'-\mathbf{r}_i^b)\right\rangle \\
&\quad + \left\langle \sum_{i=1}^{N}\sum_{j\neq i}^{N-1}\delta(\mathbf{r}-\mathbf{r}_i^a)\delta(\mathbf{r}'-\mathbf{r}_j^b)\right\rangle - \rho^2
\end{aligned}
\tag{2.43}
$$

The first term in the right hand side is separated into two terms depending on $a = b$ or $a \neq b$ as,

$$
\begin{aligned}
\left\langle \sum_{i=1}^{N}\delta(\mathbf{r}-\mathbf{r}_i^a)\delta(\mathbf{r}'-\mathbf{r}_i^b)\right\rangle &= \left\langle \sum_{i=1}^{N}\delta(\mathbf{r}-\mathbf{r}_i^a)\delta(\mathbf{r}'-\mathbf{r}_i^a)\right\rangle \\
&\quad + \left\langle \sum_{i=1}^{N}\delta(\mathbf{r}-\mathbf{r}_i^a)\delta(\mathbf{r}'-\mathbf{r}_i^b)\right\rangle \\
&= \rho\delta_{ab}\delta(\mathbf{r}-\mathbf{r}') + \rho(1-\delta_{ab})s_{ab}(\mathbf{r},\mathbf{r}')
\end{aligned}
\tag{2.44}
$$

where $s_{ab}(\mathbf{r},\mathbf{r}')$ is the *intra*-molecular correlation function defined by

$$
s_{ab}(\mathbf{r},\mathbf{r}') \equiv \left\langle \sum_{i=1}^{N}\delta(\mathbf{r}-\mathbf{r}_i^a)\delta(\mathbf{r}'-\mathbf{r}_i^b)\right\rangle. \tag{2.45}
$$

The second and third terms represent the intermolecular correlation,

$$
\left\langle \sum_{i=1}^{N}\delta(\mathbf{r}-\mathbf{r}_i^a)\sum_{j\neq i}^{N-1}\delta(\mathbf{r}'-\mathbf{r}_j^b)\right\rangle - \rho^2 \equiv \rho^2 h'_{ab}(\mathbf{r},\mathbf{r}') \tag{2.46}
$$

So, it may be understood that the pair correlation function for molecular liquids consists of two contributions, one from the intra-molecular interactions or *bonds*, $s_{ab}(\mathbf{r},\mathbf{r}')$, and the other from the intermolecular contributions, $h'_{ab}(\mathbf{r},\mathbf{r}')$. It should be noted that $h'_{ab}(\mathbf{r},\mathbf{r}')$ is not independent from $s_{ab}(\mathbf{r},\mathbf{r}')$, because the correlation between two atoms, each belonging to different molecules, is influenced by the other atoms in the same molecule. (Remember the integration variables in Eq. (2.41) include the atomic coordinates of the molecule concerned with.)

In order to express such a situation, we define a *hyper-vertex* function by,

$$\omega_{ab}(r) = \rho\delta_{ab}\delta(r) + \rho(1 - \delta_{ab})s_{ab}(r) \tag{2.47}$$

where δ_{ab} denotes the Kronecker delta function, and $s_{ab}(r)$ represents the probability distribution of an atom b around an atom a in the same molecule. If the molecule is rigid, $s_{ab}(r)$ can be expressed by the following equation in terms of the Dirac delta function as,

$$s_{ab}(r) = \frac{1}{4\pi L_{ab}}\delta(r - L_{ab}), \tag{2.48}$$

where L_{ab} denotes the *bond-length* between atoms a and b in the same molecule. So, if one chooses the first term in Eq. (2.47), it signifies just a position of a single atom, while the second term represents the two atoms connected by a chemical bond.

Using the definition of the hypervertex, Eq. (2.47), the chain sum for molecular liquids may be written in the form,

$$\begin{aligned}\rho\mathbf{h}\rho &= C[\mathbf{c}|\omega] \\ &= \omega * \mathbf{c} * \omega + \omega * \mathbf{c} * \omega * \mathbf{c} * \omega + \cdots\end{aligned} \tag{2.49}$$

in which all the variables ρ, $\mathbf{h}$, $\mathbf{c}$, and ω are matrices, elements of which are labeled by atoms in a molecule, O, H, H, in the case of water. Accordingly, the symbol "$*$" used for the convolution integral also represents the matrix product. In the Fourier space, Eq. (2.49) may be written as an infinite series of *products* of the two functions, which can be analytically summed up to give,

$$\begin{aligned}\rho h(\mathbf{k})\rho &= \omega(\mathbf{k})\mathbf{c}(\mathbf{k})\omega(\mathbf{k}) + \omega(\mathbf{k})\mathbf{c}(\mathbf{k})\omega(\mathbf{k})\mathbf{c}(\mathbf{k})\omega(\mathbf{k}) + \cdots \\ &= \frac{\omega(\mathbf{k})\mathbf{c}(\mathbf{k})\omega(\mathbf{k})}{1 - \omega(\mathbf{k})\mathbf{c}(\mathbf{k})}\end{aligned} \tag{2.50}$$

The equation is referred to as the Reference Interaction-Site Model (RISM) equation derived by Chandler and Andersen, which has become a great break-through in the statistical mechanics of molecular liquids (Chandler and Andersen 1972).

The theory, however, is not enough to be applied to *water* as is, because the interaction between atoms in water molecules includes the electrostatic interaction between partial charges put on the atoms (O, H, H). It is the theory developed by Hirata and Rossky that has solved the problem by means of the renormalization of the Coulomb interaction, similar to that developed for the electrolyte solution by Friedman and his coworkers (see Sect. 2.7) (Hirata and Rossky 1981; Friedman 1985).

2.8 Density Fluctuations in Water (XRISM Theory)

The interaction between a pair of atoms of water molecules is expressed by,

$$u_{ab}(r) = u^*_{ab}(r) + \frac{z_a z_b}{r}, \tag{2.51}$$

where $u^*_{ab}(r)$ denotes the short-range interaction between a pair of atoms, represented popularly by the Lennard–Jones interaction, and the second term expresses the Coulomb interaction between the partial charges, z_a and z_b, on the O and/or H atoms.

Now, let us split the direct correlation function as,

$$c_{ab}(r) = c^s_{ab}(r) + \phi_{ab}(r) \tag{2.52}$$

where $c^s_{ab}(r)$ and $\phi_{ab}(r)$ are defined by,

$$c^s_{ab}(r) \equiv c_{ab}(r) - \phi_{ab}(r), \quad \phi_{ab}(r) \equiv \frac{1}{k_B T}\frac{z_a z_b}{r} \tag{2.53}$$

Substituting those definitions in the chain-sum expression of the RISM equation, Eq. (2.49), one finds the following expression for the RISM equation,

$$\rho \mathbf{h} \rho = C\left[\mathbf{c}^s|\Omega\right] + \mathbf{Q}, \tag{2.54}$$

where $\mathbf{Q}$ is the Coulomb interactions between the partial charges of atoms in water molecules, renormalized with the hypervertex function ω_{ab} defined by Eq. (2.47),

$$\mathbf{Q} \equiv C[\phi|\omega] = C'[\phi|\omega] + \phi \tag{2.55}$$

and Ω is another vertex function defined by,

$$\Omega \equiv \omega + \rho C[\phi|\omega] = \omega + \rho \mathbf{Q} \rho. \tag{2.56}$$

It can be readily proved that $Q_{ab}(r)$ decays asymptotically as $\lim_{r\to\infty} Q_{ab}(r) \to \sim 1/r^6$. Thus, the problem concerning the divergence of Coulomb integral was solved. The equation is referred to as Extended RISM (XRISM) equation (Hirata and Rossky 1972).

In order to solve the equation, it is necessary to find a closure relation, such as the HNC equation. For that purpose, let's define another function τ by,

$$\tau_{ab} \equiv C\left[c^s|\Omega\right]_{ab} - c^s_{ab}, \tag{2.57}$$

that is the indirect correlation function from which the long-range character is removed. Then, the HNC closure for the XRISM theory is expressed as,

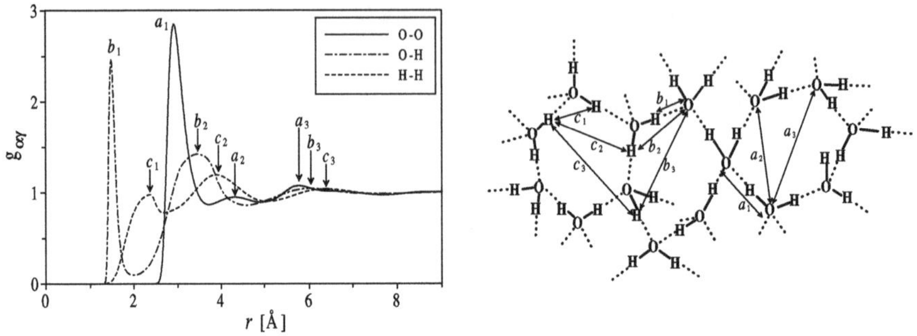

Fig. 2.6 The site–site pair correlation function of water calculated by the XRISM theory: a_1, b_1, c_1, etc. correspond to the distance between atoms, depicted in the cartoon of water molecules, hydrogen-bonded between one and another. The hydrogen-bonds are depicted by dotted-lines

$$c^{s}_{ab}(r) = \exp\left[-u^{*}_{ab}(r) + Q_{ab}(r) + \tau_{ab}(r)\right] - 1 - \left[Q_{ab}(r) + \tau_{ab}(r)\right]. \tag{2.58}$$

In Eq. (2.58), the Coulomb interaction is renormalized into $Q_{ab}(r)$ which behaves like $\sim 1/r^6$.

Equations (2.54) and (2.58) have been applied to water by Pettitt and Rossky to produce the pair correlation function depicted in Fig. 2.6 (Pettitt and Rossky 1982). The three important characteristics of the pair correlation functions of water, identified by the X-ray diffraction and the neutron scattering experiments, are well reproduced: (i) appearance of the sharp peak (b_1) at $r \sim 1.6\sigma$ in the O–H correlation function, which is assigned to the hydrogen-bond between a pair of water molecules; (ii) the sharpening of the first peak in the O–O correlation function, which is a manifestation of the tetrahedral coordination due to the hydrogen bond; (iii) appearance of small peak between the first and second peaks in the O–O correlation function, which is attributed to the tetrahedral coordination due to the hydrogen-bond (Thiessen and Narten 1982). These peaks have been referred to as the "signature" of water structure, since those peaks represent the tetrahedral coordination of water molecules due to the hydrogen-bonds. A schematic view of the hydrogen-bonded structure is depicted in the right-hand panel in Fig. 2.6.

2.9 Density Fluctuation of Water Around a Biomolecule

If a biomolecule such as protein is fixed in aqueous solution, the density of water around the biomolecule is highly heterogeneous. Therefore, it is the first moment or the average density that characterizes the structure of water around the biomolecule. A theory to describe the structure of water in inhomogeneous field around a biomolecule is the 3D-RISM/RISM theory (Beglov and Roux 1997; Kovalenko and Hirata 1998).

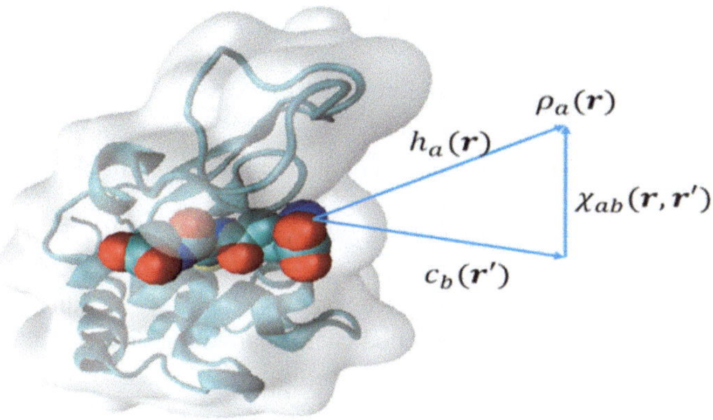

Fig. 2.7 Inhomogeneous density distribution of solvent around a protein

Here, we briefly review the 3D-RISM/RISM theory, focusing on its physics, not its formulation. And then, we outline the route to find the free energy surface, and its first and second derivatives with respect to the atomic coordinate of protein.

Let us consider the average density of solvent molecules at a position around a solute molecule (Fig. 2.7). When the position is far from the solute molecule so as to be regarded as the bulk, the density will be constant which is the same as in the pure solvent. On the other hand, when it is next to a solute, the density will be "perturbed" significantly by the field due to the solute molecule, and is different from that in the bulk, depending on the strength of the perturbation. Physical meaning of the 3D-RISM equation can be readily understood in terms of a *non-linear* response theory as follows.

Let us denote the average density of solvent atom j at the bulk, the density nearby solute, and the density response to the perturbation as ρ_a, $\rho_a(\mathbf{r})$, and $\Delta\rho_a(\mathbf{r})$, respectively, where the subscript a specifies an atom of water. Then, statement above can be expressed as,

$$\rho_a(\mathbf{r}) = \rho_a + \Delta\rho_a(\mathbf{r}) \tag{2.59}$$

The density response to the perturbation can be expressed in terms of a non-linear response theory as

$$\Delta\rho_a(\mathbf{r}) = \sum_b \int \rho_a \chi_{ab}(\mathbf{r}, \mathbf{r}')c_b(\mathbf{r}')d\mathbf{r}' \tag{2.60}$$

where the $c_b(\mathbf{r})$ is a *non-linear* perturbation due to the solute molecule, and $\chi_{ab}(\mathbf{r}, \mathbf{r}')$ is a response function. (The reason why we call it "non-linear" will be clarified in a moment.) The equation can be viewed as the 3D-RISM equation with identifying $c_b(\mathbf{r})$ as the direct correlation function and $\chi_{ab}(\mathbf{r}, \mathbf{r}')$ as the correlation function of density fluctuation in bulk solvent, that is,

$$\rho^2 \chi_{ab}(\mathbf{r}, \mathbf{r}') = \left\langle \delta\rho_a^{(0)}(\mathbf{r}) \delta\rho_b^{(0)}(\mathbf{r}') \right\rangle \tag{2.61}$$

where $\delta\rho_a^{(0)}(\mathbf{r})$ is the density fluctuation of atom a in the pure liquid defined by

$$\delta\rho_a^{(0)}(\mathbf{r}) = \rho_a^{(0)}(\mathbf{r}) - \rho_a. \tag{2.62}$$

Several approximate equations have been devised for the direct correlation functions. For examples, the hypernetted-chain approximation reads,

$$c_a(\mathbf{r}) = \exp[-u_a(\mathbf{r})/k_{\mathrm{B}}T + h_a(\mathbf{r}) - c_a(\mathbf{r})] - 1 - [h_a(\mathbf{r}) - c_a(\mathbf{r})] \tag{2.63}$$

In the expression, $u_a(\mathbf{r})$ is the direct interaction potential exerted on the solvent atom, a, from the solute molecule, and $h_a(\mathbf{r})$ is the density fluctuation in solvent at position $\mathbf{r}$, normalized by the bulk density, namely,

$$h_a(\mathbf{r}) = \Delta\rho_a(\mathbf{r})/\rho \tag{2.64}$$

The three-dimensional distribution function is defined from $h_a(\mathbf{r})$ by

$$g_a(\mathbf{r}) \equiv h_a(\mathbf{r}) + 1 \tag{2.65}$$

It is not only the direct interaction $u_a(\mathbf{r})$ with solute that perturbs the density of solvent at a certain position, but also that from solvent molecules at the other positions, whose density is also perturbed by the existence of the same solute. Such "indirect" perturbations are *renormalized* into the terms including $(h_a(\mathbf{r}) - c_a(\mathbf{r}))$. Such renormalization makes the perturbation highly *nonlinear*.

The procedure of solving the equations consists of two steps. In the first step, we solve the XRISM equation Eq. (2.49) for $h_{ab}^{(0)}(r)$ of solvent, or a mixture of solvents in cases of solutions. Then, we solve the 3D-RISM equation, Eq. (2.60), for $h_a(\mathbf{r})$ of a protein-solvent (solution) system, plugging $h_{ab}^{(0)}(r)$ for the solvent into Eq. (2.60), which has been calculated in the first step.

$h(\mathbf{r})$ or $g(\mathbf{r})$ thus obtained is the three-dimensional (3D) distribution of solvent molecules around a protein in terms of the interaction-site representation of solvent, or a mixture of solvents. In the following, let us briefly review a result of calculation based on the RISM/3D-RISM theory, applied to a protein called "hen egg-white lysozyme" in water (Imai et al. 2005a, b).

2.10 Water Molecules *Recognized* by a Biomolecule

Imai et al. carried out the RISM/3D-RISM calculation for a hen egg-white lysozyme immersed in water and obtained the 3D-distribution function ($g(\mathbf{r})$) of oxygen and hydrogen of water molecules around and inside the protein (Imai et al. 2005a, b). The protein is known to have a cavity composed of the residues from Tyr53 to Ile58 and from Ala82 to Ser91, in which four water molecules have been located by means of the X-ray diffraction measurement (Kuroki and Yutani 1998). In the calculation, those water molecules are not included explicitly.

In Fig. 2.8, depicted by green surfaces or spots are $g(\mathbf{r})$ of oxygen atoms of water molecules using an iso-surface representation, which is very similar to the electron density map obtained from the X-ray crystallography. The authors have drawn $g(\mathbf{r})$ which is greater than a threshold value. The right, center, and left figures correspond, respectively, to $g(\mathbf{r}) > 2.0$, $g(\mathbf{r}) > 4.0$, and $g(\mathbf{r}) > 8.0$. Since $g(\mathbf{r})$ is unity in the bulk, the left figure indicates that the probability of finding those water molecules at the surface is more than twice as larger compared to the bulk water. As such, the water molecules depicted in the right figure has eight times higher probability to be found than those in the bulk. The water molecules are those bound firmly to some particular atom of the protein due to, for examples, the hydrogen bonds, and they are quite rare as one can see from the figure. In this sense, the threshold values play the role of the "temperature" in the X-ray diffraction measurement: if you lower the temperature, you can observe more water molecules which have weaker interaction with protein.

Shown in Fig. 2.9 are the distributions of water molecules *inside a cavity* which is formed by residues from Tyr53 to Iso58 and from Ala82 to Ser91. The left picture (a) in Fig. 2.9 shows the isosurfaces of $g(\mathbf{r}) > 8$ for water-oxygen (green) and hydrogen (pink) in the cavity. In the figure, only the surrounding residues are displayed, except for Ala82 and Lys83 which are located in the front side. There are four distinct peaks of water oxygen and seven distinct peaks of water hydrogen in the cavity. The spots colored by green and pink indicate water oxygen and hydrogen, respectively. From

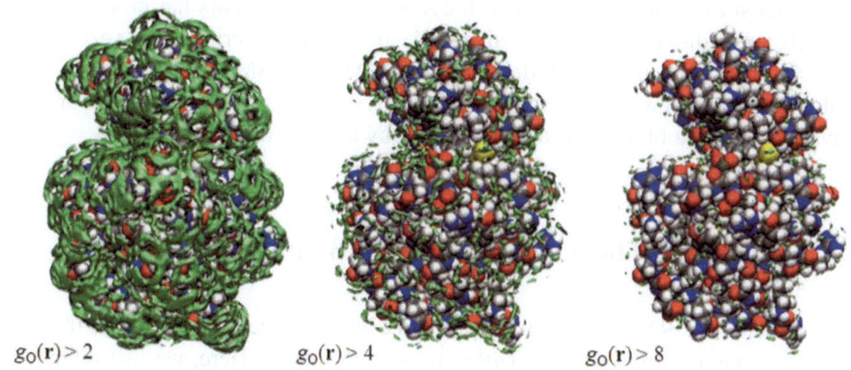

Fig. 2.8 3D-distribution $g(r)$ of water around a protein (lysozyme)

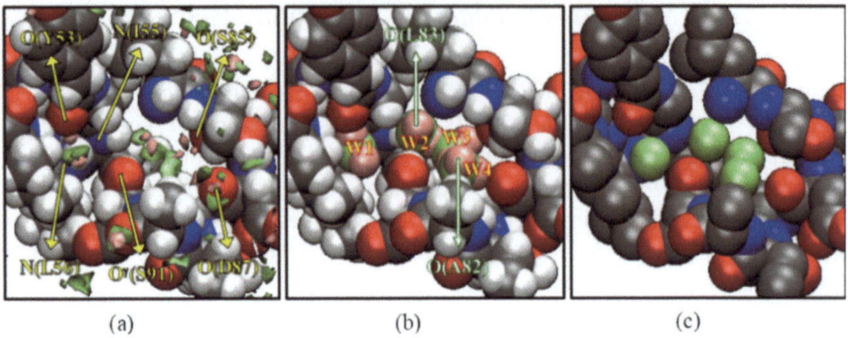

Fig. 2.9 The 3D-distribution $g(\mathbf{r})$ of water inside the active site of protein: green surface, oxygen; purple surface, hydrogen

the isosurface plots, Imai et al. have reconstructed the most probable model of the hydration structure. It is shown in Fig. 2.9b, where the four water molecules are numbered in the order from the left. Water 1 is hydrogen-bonding to the main-chain oxygen of Tyr53 and the main-chain nitrogen of Leu56. Water 2 forms hydrogen bonds to the main-chain nitrogen of Iso56 and the main-chain oxygen of Leu83, which is not drawn in the figure. Water 3 and 4 also form hydrogen bonds with protein sites, the former to the main-chain oxygen of Ser85 and the latter to the main-chain oxygens of Ala82 (not displayed) and of Asp87. There is also a hydrogen bond network among Water 2, 3, and 4. The peak of the hydrogen between Water 3 and 4 does not appear in the figure because it is slightly less than 8, which means the hydrogen bond is weaker or looser than the other hydrogen-bonding interactions. Although the hydroxyl group of Ser91 is located in the center of the four water molecules, it makes only weak interactions with them.

The hydration structure obtained by the RISM/3D-RISM theory is compared with crystallographic water sites of X-ray structure (Kuroki and Yutani 1998). The crystallographic water molecules in the cavity are depicted in the right of Fig. 2.9c showing four water sites in the cavity, much as the RISM/3D-RISM theory has probed. Moreover, the water distributions obtained from the theory and experiment are very similar to each other. Thus, it is concluded that the RISM/3D-RISM theory can predict the water-binding sites with great success. This was the first in the history of theoretical physics to probe a little molecule *recognized* by a cavity of a biomolecule.

2.11 Solvation Thermodynamics of a Biomolecule in Water

Provided that the 3D-distribution $g(\boldsymbol{r})$ around a biomolecule is obtained, the thermodynamic quantities related to *solvation* may be evaluated. Here, we just refer four thermodynamic quantities that play crucial roles in analyzing the conformational stability of biomolecules in water: the chemical potential, the solvation energy, the isothermal compressibility, and the partial molar volume.

Chemical Potential

The RISM formula of the chemical potential, or the solvation free energy, was derived by Singer and Chandler, with the HNC closure (Singer and Chandler 1985). The equation can be *formally* generalized to a biomolecule in water as

$$\Delta\mu = \rho k_B T \sum_a \int \left[\frac{1}{2} h_a(\mathbf{r})^2 - c_a(\mathbf{r}) - \frac{1}{2} h_a(\mathbf{r}) c_a(\mathbf{r}) \right] d\mathbf{r}, \tag{2.66}$$

where $h_a(\mathbf{r})$ are the density fluctuation of the atom a of solvent molecules around the biomolecule, defined by Eq. (2.64), and $c_a(\mathbf{r})$ is the direct correlation function defined by Eq. (2.63) (Kovalenko and Hirata 1998).

Solvation Energy

The solvation energy is calculated from the chemical potential using the Gibbs–Helmholtz formula in thermodynamics.

$$\Delta E = \left(\frac{\partial(\Delta\mu/T)}{\partial(1/T)} \right)_{N,V} \tag{2.67}$$

Isothermal Compressibility Pf Solvent

The isothermal compressibility χ_T of solvent is evaluated by means of the XRISM theory as,

$$k_B T \chi_T = \left[\rho - \rho^2 \sum_{a=1}^{n} \sum_{b=1}^{n} \int c_{ab}(\mathbf{r}) d\mathbf{r} \right]^{-1} \tag{2.68}$$

where $c_{ab}(\mathbf{r})$ denotes the direct correlation function between a pair of atoms a, b in solvent molecules, and ρ is the number density of solvent (Imai et al. 2005a, b).

Partial Molar Volume

The partial molar volume of a biomolecule in solvent is calculated by the following formula,

$$\overline{V} = k_B T \chi_T \left[1 - \rho \sum_a \int c_a(\mathbf{r}) d\mathbf{r} \right] \tag{2.69}$$

where a labels atoms of solvent molecules, ρ is the density of solvent, and χ_T is the isothermal compressibility of solvent.

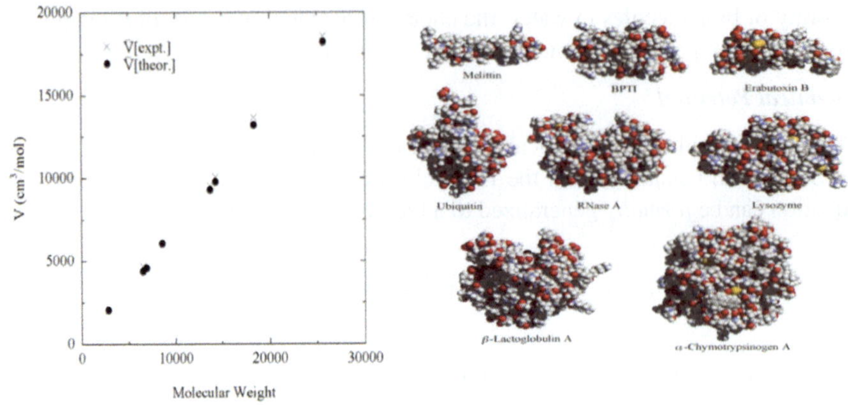

Fig. 2.10 The partial molar volume of a variety of protein in water

The partial molar volume has been calculated by Imai et al. based on Eq. (2.69) for a variety of proteins. The result is exhibited in Fig. 2.10. The partial molar volume is a thermodynamic variable that plays a crucial role in the structural change of protein, induced by *pressure* (Imai et al. 2000, 2005a, b; Kirkwood and Buff 1951).

Appendix: On the Federov Correction to the Solvation Free Energy

In the chapter, we have reviewed the RISM and 3D-RISM theories to treat the density fluctuation of water around biomolecules. It has been shown that the theory is quite successful to explain solvation structure of a biomolecule in aqueous solution. The theory also explains the partial molar volume of biomolecules in water, one of the thermodynamic quantities, quite accurately.

Despite the remarkable success, the theories are notorious concerning the quantitative aspect for reproducing experimental results of the solvation free energy associated with the process dissolving molecules in solution, especially for the hydrophobic contribution. Many remedies proposed to cure the trouble, which are almost thoroughly concerned with the approximation inherent to the RISM theories, such as "bridge functionals" and/or "improper diagrams", have not been very successful, or just in partial success.

Recently, a method to correct the RISM error has been proposed by Fedorov and his coworkers, which is based entirely on an empirical consideration (Ratkova et al. 2010). Let us refer the correction to "Federov correction". The importance of their finding lies in the fact that the correction term is proportional to the *partial molar volume* of solute in solutions, one of thermodynamic quantities. The observation suggests strongly that the correction may not be sensitive to the detailed microscopic structure of solvation, with which such quantities as the "bridge functionals" and/or "improper diagrams" are concerned.

There is an important observation which may suggest a source of the discrepancy between the theoretical and empirical results concerning the solvation free energy. The observation is the partial molar volume of protein calculated based on the 3D-RISM/RISM theory combined with the Kirkwood–Buff theory. According to the results by Imai et al., the partial molar volume of many protein molecules in aqueous solutions, plotted against their molecular weight, shows quantitative agreement with corresponding experimental data (see Fig. 2.10). The essential ingredient of the Kirkwood–Buff theory lies in the transformation from the (V, T) to (P, T) ensembles using the Gibbs–Duhem theorem. The fact that the theory is successful to explain the partial molar volume of protein suggests strongly that the error in the RISM theory originates, at least partly, from the difference in the ensembles between the theory and the experimental set up.

Friedman's Correction: Actually, such an inconsistency between thermodynamic ensembles has been realized earlier by several scientists in the field of the statistical mechanics of liquid and solution. The theoretical calculations are carried out on a system of constant volume (*molarity* scale and Helmholtz free energy, "*F*"), while most experimental data are measured in terms of the *molality* scale and Gibbs free energy, "*G*", and activity of the solvent varies with concentration. It is, therefore, necessary to correct one of those to make comparison possible. Such a correction has been done by Poirier, who corrected the theoretical results to the experimental ones: the theoretical results of the Helmholtz free energy to the Gibbs free energy (Poirier 1953). On the other hand, Friedman has made the correction in opposite direction, namely, from experimental results to theoretical ones (Friedman 1960; Hirata and Arakawa 1975). Ignoring all terms after the second term of the correction, one gets,

$$f = g + \frac{1}{2}RT\ln\left[\frac{V(m,0)}{V(0,0)}\right], \tag{2.70}$$

where f is the Helmholtz free energy per concentration measured by molarity (mol/L) scale, and g is the Gives free energy per concentration measured by molality (mol/kg) scale, respectively. $V(m, 0)$ the volume of an m molal solution having 1 kg of solvent, and $V(0, 0)$ the volume of 1 kg of solvent, both at 0 pressure. g is related to the osmotic coefficient (ϕ) and the mean activity coefficient ($\gamma_\pm$) by the equation,

$$g = RTm\left[1 - \phi + \ln\gamma_\pm\right]. \tag{2.71}$$

The chemical potential of species i in (V, T)- and (P, T)-ensemble are, respectively, defined by,

$$\mu_\mathrm{H} = \left(\frac{\partial f}{\partial c_i}\right)_{V,T} \quad \text{and} \quad \mu_\mathrm{G} = \left(\frac{\partial g}{\partial m_i}\right)_{P,T}. \tag{2.72}$$

Therefore, there is a difference between the two chemical potentials obtained from the two ensembles (V, T) and (P, T), and the difference is essentially,

$$\sim \frac{1}{V}\left(\frac{\partial V}{\partial m_i}\right)_{P,T}. \tag{2.73}$$

The quantity is nothing but the *partial molar volume* properly normalized for the unit volume.

The author believes that this is the correction discovered empirically by Fedorov and his coworkers.

Heuristic Explanation of the Federov Correction: In the case of experiment, N and V are changed independently, and the density of the system is defined by their ratio, or N/V. However, the setup of ensembles in the statistical mechanics may not meat such a premise. The statistical mechanics formulation of liquid state requires that the system should be in thermodynamic limit, $N \rightarrow \infty$ and $V \rightarrow \infty$, and the ratio N/V or the density should be constant. Due to the restriction, V and N cannot be independent. In another word, when a species is added into the system, the volume of the system should be changed accordingly in order to keep the density constant, or if the volume is kept constant, the density should be changed. In the conventional formulation of the statistical mechanics, it is somewhat obscured by taking the thermodynamic limit, $V \rightarrow \infty$. Unfortunately, it is not possible to reformulate the theory by removing one of those restrictions. Therefore, we should take an alternative way to estimate a correction to the error originated from the contradiction between the theory and experiments. The simplest way of the correction is to remove the restriction concerning the volume, and let the volume change upon the dissolving process of solute. The correction then should be the work required to compress (or expand) the system to the original volume against the internal (or external) pressure.

The increase or decrease of volume upon adding a species in the system is nothing but the partial molar volume. Suppose the partial molar volume is positive, then the correction should be the work required to compress the system from $V + \overline{V}_j$ to V against the internal pressure of liquid, where V and $\overline{V}_j$ are the volume of the system to be constant and the partial molar volume of species j. The internal pressure depends on the volume, and sharply increases upon compression. Therefore, the correction should be expressed by the following equation.

$$\Delta\mu_{\text{correct}} = \int_{V}^{V+\overline{V}_j} p(V)\mathrm{d}V$$

In the equation, the pressure at $V + \overline{V}_j$ should be one atmosphere at the standard state, since the internal pressure should be balanced with external pressure. The internal pressure increases upon compression to reach $p(V)$ at the volume originally set up for the ensemble.

The correction is roughly proportional to the partial molar volume as can be understood by approximating $p(V)$ by its average $\langle p \rangle$ over the volume change, that leads,

$$\Delta\mu_{\text{correct}} \sim \langle p \rangle \overline{V}_j$$

This may be the rational physics of the empirical correction found by Federov and coworker, which is proportional to the partial molar volume of solute. The physics also explains why the solvation free energy calculated by the RISM/3D-RISM theory is quite different from the experimental results, while those of the partial molar volume nicely agrees with corresponding experiments.

References

Beglov D, Roux B (1997) J Phys Chem B 101:7821

Chandler D, Andersen HC (1972) Optimized cluster expansions for classical fluids. II. Theory of molecular liquids. J Chem Phys 57:1930–1931

Debye P, Huckel E (1923) De la theorie des electrolytes. I. Abaissement du point de congelation et phenomenes associes. Phys Z 24(9):185–206

Friedman HL (1960) J Chem Phys 32:1351

Friedman HL (1985) A course in statistical mechanics. Prentice-Hall, New Jersey

Hansen JP, McDonald IR (1986) Theory of simple liquids. Academic Press, London

Harashima A (1954) Theory of liquids. Iwanami, Tokyo (written in Japanese)

Hirata F (2020) Exploring life phenomena with statistical mechanics of molecular liquids. CRC Press, New York

Hirata F, Rossky PJ (1981) An extended RISM equation for molecular polar fluids. Chem Phys Lett 329–334

Hirata F, Arakawa K (1975) The computation of the thermodynamic properties of aqueous electrolyte solutions by means of the perturbation theory of fluid. Bull Chem Soc Jpn 48:2139

Hirata F, Rossky PJ (1981) An extended RISM equation for molecular polar fluids. Chem Phys Lett 329–334

Imai T, Kinoshita M, Hirata F (2000) Theoretical study for partial molar volume of amino acids in aqueous solution: implication of ideal fluctuation volume. J Chem Phys 112:9469–9478

Imai T, Hiraoka R, Kovalenko A, Hirata F (2005a) Water molecules in a protein cavity detected by a statistical-mechanical theory. J Am Chem Soc 127:15334–15335

Imai T, Kovalenko A, Hirata F (2005b) Partial molar volume of proteins studied by the three-dimensional reference interaction site model theory. J Phys Chem B 109:6658–6665

Kirkwood JG, Buff FP (1951) The statistical mechanical theory of solutions. J Chem Phys 19:774–777

Kovalenko A, Hirata F (1998) Three-dimensional density profiles of water in contact with a solute of arbitrary shape: a RISM approach. Chem Phys Lett 290:237–244

Kuroki R, Yutani K (1998) Structural and thermodynamic responses of mutations at a Ca^{2+} binding site engineered into human lysozyme. J Biol Chem 273:34310–34315

McQuarrie DA (1976) Statistical mechanics. Harper and Row, New York

Ornstein LS, Zernike F (1914) Accidental deviations of density and opalescence at the critical point of a single substance. Proc R Netherlands Acad Arts Sci 17:793–806

Pettitt MB, Rossky PJ (1982) Integral equation predictions of liquid-state structure for water like intermolecular potentials. J Chem Phys 77:1451–1457

Poirier JC (1953) J Chem Phys 21(965):972
Ratkova EL, Chuev GN, Sergiievskyi FMV (2010) An accurate prediction of hydration free energies by combination of molecular integral equation theory with structural descriptors. J Phys Chem B 114:12068–12079
Rossky PJ, Dale WDT (1980) Generalized recursive solutions to Ornstein–Zernike integral-equation. J Chem Phys 73:2457–2464
Singer SJ, Chandler D (1985) Free-energy functions in the extended RISM approximation. Mol Phys 55:621–625
Thiessen WE, Narten AH (1982) Neutron diffraction study of light and heavy water mixtures at 25 °C. J Chem Phys 77:2655–2662

Chapter 3
Generalized Langevin Theory and Density Fluctuations in Water

The present chapter is devoted to explain the generalized Langevin theory and its application to water. The GLT has been originally formulated for a quantum–mechanical system by H. Mori, and it has been reformulated for the classical fluid by H. L. Friedman (Mori 1958, 1962, 1965; Friedman 1985). The theory was applied to the interaction-site model of water by the author in 1992 (Hirata 1992).

In Sects. 3.1 and 3.2, we briefly outline the statistical-mechanics theory referred to as the *generalized Langevin theory* (GLT). The GLT has a theoretical framework that is similar to the classical or phenomenological Langevin theory (PLT), both having the frictional and random-force terms in the respective equation of motion. The well-regarded difference between the two theories lies in the description concerning the origin of those forces. The PLT describes the forces in terms of phenomenological theory such as the hydrodynamics, while the GLT derives those forces from the first principle in the phase space as a consequence of projecting all the variables in the phase space onto the dynamic variable of concern. There is another difference between the two theories, which is essential for describing thermal fluctuations of a system. Any fluctuation taking place in solution, due to the thermal motion, should restore its equilibrium state. Such a force to restore the equilibrium state is not included in the PLT formalism. On the other hand, the restoring force is naturally included in the GLT equation as a consequence of the projection of all mechanical variables onto the dynamic variables.

In Sect. 3.3, we review the GLT study of water, in which the generalized Langevin equation for water is derived. In Sect. 3.4, the application of the theory to the temporal density fluctuations in water is presented. Section 3.5 is devoted to a dispersion-relation analysis of the space–time correlation function in water. Based on the analysis, it is concluded that the density fluctuation of water is characterized by three collective modes, one acoustic and two optical modes.

F. Hirata, *Molecular Theory of Fluctuation in Life Phenomena*,
https://doi.org/10.1007/978-981-96-3364-7_3

3.1 Equation of Motion in Phase Space

We consider a system consisting of N molecules: $N \sim 10^{23}$. We assume that the particles are all classical, meaning that each molecule follows the Newtonian equation. A dynamical state of such a system is defined uniquely by a point in the space called the *phase space*. The phase space is the coordinates space spanned by 6N variables. The 3N variables of the coordinate system are the spatial coordinates: The spatial coordinate of each molecule is denoted by q_x, q_y, q_z. The other 3N variables are assigned to x, y, z-components of the momentum, p_x, p_y, p_z. It is worthwhile to be noted that the two variables of a molecule in an quantum system is not independent due to the Heisenberg *uncertainty principle*, or $\Delta x \Delta p_x \sim h$. The 6$N$ dimensional space is referred to as "phase space". We refer to a point in the phase space as "phase point" or "representative point", and denote it by $\Gamma_N = \{q_1, q_2, \ldots, q_{3N}; \ p_1, p_2, \ldots, p_{3N}\}$.

The Hamiltonian of the system is defined by

$$H = \sum_i^{3N} p_i^2/2m_i + U_N(q_1, q_2, \ldots, q_{3N}) \tag{3.1}$$

The derivative of Eq. (3.1) with respect to the momentum and spatial coordinates leads the following equations of motion,

$$\frac{\partial H}{\partial p_i} = \dot{q}_i, \quad \frac{\partial H}{\partial q_i} = -\dot{p}_i \tag{3.2}$$

It is worthwhile to note that all the mechanical properties, dynamical as well as static, are determined thoroughly by the phase point Γ_N. Such variables in phase space are referred to as "dynamic variables". The Hamiltonian defined by Eq. (3.1) and its elements, the kinetic and potential energies are typical examples of the dynamic variables. The density field defined by Eq. (2.4) in Chap. 2 is another example of the dynamic variables.

Let us consider temporal change of a dynamic variable, A, in the phase space. It may be expressed by a sum of two terms as,

$$\frac{\mathrm{d}A}{\mathrm{d}t} = \left(\frac{\partial A}{\partial t}\right)_{\Gamma_N} + \sum_i \left(\frac{\partial A}{\partial q_i}\dot{q}_i + \frac{\partial A}{\partial p_i}\dot{p}_i\right) \tag{3.3}$$

The first term in the right-hand-side represents the temporal change of A at the fixed phase point Γ_N. The term vanishes by definition, because a dynamic variable depends only on the phase Γ_N. That is,

$$\left(\frac{\partial A}{\partial t}\right)_{\Gamma_N} = 0 \tag{3.4}$$

Therefore, one finds

$$\frac{dA}{dt} = -\sum_i \left(\frac{\partial H}{\partial q_i} \frac{\partial A}{\partial p_i} - \frac{\partial H}{\partial p_i} \frac{\partial A}{\partial q_i} \right) \equiv \hat{L}' A \tag{3.5}$$

where we have defined the Liouville operator $\hat{L}'$ by

$$\hat{L}' = -\sum_i \left(\frac{\partial H}{\partial q_i} \frac{\partial}{\partial p_i} - \frac{\partial H}{\partial p_i} \frac{\partial}{\partial q_i} \right). \tag{3.6}$$

The operator may be expressed in more explicit formula as

$$\hat{L}' = \sum_i^{3N} \left(\frac{p_i}{m_i} \frac{\partial}{\partial q_i} + F_i \frac{\partial}{\partial p_i} \right). \tag{3.7}$$

In the equation, $F_i = -dH/dq_i = -dU/dq_i$. Equation (3.5) is referred to as *Liouville equation of motion*.

Liouville Equation and Theorem: Let us consider an ensemble of replicas of a system including N molecules. Each replica of the ensemble corresponds to a point in the phase space, and the ensemble makes a distribution that changes with time. Denoting the distribution of phase points by f_N, its temporal change is expressed by

$$\begin{aligned} \frac{df_N}{dt} &= \left(\frac{\partial f_N}{\partial t} \right)_{\Gamma_N} + \sum_{i=1}^{3N} \left(\frac{\partial f_N}{\partial q_i} \dot{q}_i + \frac{\partial f_N}{\partial p_i} \dot{p}_i \right) \\ &= \left(\frac{\partial f_N}{\partial t} \right)_{\Gamma_N} + \hat{L}' f_N \end{aligned} \tag{3.8}$$

The left-hand-side of Eq. (3.8) vanishes, or,

$$\frac{df_N}{dt} = 0. \tag{3.9}$$

The relation is referred to as "Liouville theorem". The proof of the theorem is provided in a moment. On the other hand, the first term in the right-hand-side should not vanish, since the population of replicas at a phase point varies with time. In the following, we prove the Liouville theorem, or Eq. (3.9).

The distribution of representative points in the phase space can be liken to the density of fluid in which particles are in free motion. The continuity equation of such a fluid system may be written as,

$$\left(\frac{\partial f_N}{\partial t} \right)_{\Gamma_N} = \nabla_{\Gamma_N} \cdot \mathbf{j}_{\Gamma_N}. \tag{3.10}$$

In the equation, ∇_{Γ_N} represents the vector of the differential operator in the phase space, and $\mathbf{j}_{\Gamma_N}$ denotes the *flux* of phase points, defined respectively by Eqs. (3.11) and (3.12).

$$\nabla_{\Gamma_N} \equiv \left(\frac{\partial}{\partial \mathbf{q}_1}, \frac{\partial}{\partial \mathbf{q}_2}, \ldots, \frac{\partial}{\partial \mathbf{q}_N}; \frac{\partial}{\partial \mathbf{p}_1}, \frac{\partial}{\partial \mathbf{p}_2}, \ldots, \frac{\partial}{\partial \mathbf{p}_N} \right) \tag{3.11}$$

and

$$\mathbf{j}_{\Gamma_N} \equiv (f_N \dot{\mathbf{q}}_1, f_N \dot{\mathbf{q}}_2, \ldots, f_N \dot{\mathbf{q}}_N; f_N \dot{\mathbf{p}}_1, f_N \dot{\mathbf{p}}_2, \ldots, f_N \dot{\mathbf{p}}_N) \tag{3.12}$$

The equation of continuity can be rewritten as

$$\frac{\partial f_N}{\partial t} = -\sum_i \left[\frac{\partial}{\partial q_i}(f_N \dot{q}_i) + \frac{\partial}{\partial t}(f_N \dot{p}_i) \right] \tag{3.13}$$

From Eq. (3.2), one may find the following relation

$$\frac{\partial \dot{q}_i}{\partial q_i} = \frac{\partial^2 H}{\partial q_i \partial p_i} = -\frac{\partial \dot{p}_i}{\partial q_i} \tag{3.14}$$

Substituting Eq. (3.14) into Eq. (2.13) leads to

$$\left(\frac{\partial f_N}{\partial t} \right)_{\Gamma_N} = -\sum_i \left[\frac{\partial f_N}{\partial q_i} \dot{q}_i + \frac{\partial f_N}{\partial p_i} \dot{p}_i \right] \tag{3.15}$$

Thus, the Liouville theorem, or $\mathrm{d}f_N/\mathrm{d}t = 0$, is proved.
In terms of the Liouville operator, Eq. (3.15) can be written as

$$\left(\frac{\partial f_N}{\partial t} \right)_{\Gamma_N} = -\hat{L}' f_N \tag{3.16}$$

The relation is referred to as "Liouville equation", which plays crucial roles to describe the time evolution of the dynamic variable in the phase space.

Dynamic Variables and Ensemble Average: The distribution function in equilibrium can is described by the Boltzmann distribution in the phase space,

$$f_N^{\mathrm{eq}}(\Gamma_N) = \frac{\exp[-H(\Gamma_N)/k_\mathrm{B}T]}{\int \exp[-H(\Gamma_N)/k_\mathrm{B}T]\mathrm{d}\Gamma_N}. \tag{3.17}$$

It is readily proved as follows that the distribution function in equilibrium does not depend on time,

$$\dot{f}_N^{\mathrm{eq}}(\Gamma_N) = -\hat{L}' f_N^{\mathrm{eq}}(\Gamma_N) \propto -\hat{L}' \mathrm{e}^{-H/k_\mathrm{B}T} = \left(\mathrm{e}^{-H/k_\mathrm{B}T}/k_\mathrm{B}T\right)\hat{L}' H. \tag{3.18}$$

The ensemble average of a dynamic variable D is defined by the distribution function as,

$$\langle D \rangle \equiv \int D(\Gamma_N) f_N^{\mathrm{eq}}(\Gamma_N) \mathrm{d}\Gamma_N \tag{3.19}$$

The formal solutions of the Liouville equation of motion and the Liouville equation are obtained by integrating the equations, (3.5) and (3.16) as

$$A(t) = \exp\left(i\hat{L}'\right) A(0) \tag{3.20}$$

and

$$f_N(\Gamma_N, t) = \exp\left(-i\hat{L}'\right) f_N(\Gamma_N, 0) \tag{3.21}$$

In the equations, $A(0)$ and $f_N(\Gamma_N, 0)$ represent the initial value of the dynamic variable and the distribution, respectively.

Time Correlation Functions: It is the time correlation function that plays a crucial role in the description of irreversible processes. The time correlation function is defined as the ensemble average of a product of two dynamic variables A and B at different times, $t = 0$ and $t = t$, or $D \equiv A(0)B(t)$. The time correlation function is defined based on the Liouville operator as,

$$\begin{aligned} \langle D \rangle &= \langle A(0)B(t) \rangle \\ &= \left\langle A \exp\left[\hat{L}' t\right] B \right\rangle \\ &= \int f_N^{\mathrm{eq}}(\Gamma_N) A(\Gamma_N) \exp\left[\hat{L}' t\right] B(\Gamma_N) \mathrm{d}\Gamma_N \end{aligned} \tag{3.22}$$

If the two dynamic variables A and B have the same physical properties, the time correlation function is referred to as the *auto*- or *self*-correlation function. On the other hand, it is called the *distinct*- or *joint*- or *cross*-correlation function in the case of $A \neq B$.

The time correlation function has an important property, that is, *steady-state*. The steady state is expressed in the following equation. For any interval t, the time correlation function satisfies the following relation,

$$\left\langle A(t')B(t + t') \right\rangle = \langle A(0)B(t) \rangle = \langle A(-t)B(0) \rangle. \tag{3.23}$$

The equation can be expressed in terms of the Liouville operator as (Eq. 3.19),

$$\left\langle A \exp\left[t\hat{L}'\right] B \right\rangle = \left\langle \left(\exp\left[-t\hat{L}'\right] A\right) B \right\rangle \tag{3.24}$$

In order to derive Eq. (3.24), we first derive the following relation,

$$\langle A\hat{L}'B\rangle = -\langle(\hat{L}'A)B\rangle.$$

The derivation is following.

$$\begin{aligned}
\langle A\hat{L}'B\rangle &= \int \mathrm{d}\Gamma_N f_N^{\mathrm{eq}}(\Gamma_N)A\hat{L}'B \\
&= -\int \mathrm{d}\Gamma_N f_N^{\mathrm{eq}}(\Gamma_N)A\dot{\Gamma}_N \cdot \frac{\partial}{\partial\Gamma_N}B \\
&= -\left\{\left[f_N^{\mathrm{eq}}(\Gamma_N)A\dot{\Gamma}_N B\right]_{-\infty}^{\infty} - \int \mathrm{d}\Gamma_N \dot{\Gamma}_N \cdot \frac{\partial\left\{f_N^{\mathrm{eq}}(\Gamma_N)A\right\}}{\partial\Gamma_N}B\right\} \\
&= \int \mathrm{d}\Gamma_N f_N^{\mathrm{eq}}(\Gamma_N)\left\{\dot{\Gamma}_N \cdot \frac{\partial A}{\partial\Gamma_N}\right\}B \\
&= -\int \mathrm{d}\Gamma_N f_N^{\mathrm{eq}}(\Gamma_N)\left(\hat{L}'A\right)B \\
&= -\left\langle\left(\hat{L}'A\right)B\right\rangle
\end{aligned} \tag{3.25}$$

In the derivation, the Liouville operator $\hat{L}$ is expressed in its explicit form as,

$$\hat{L}' = \dot{\Gamma}_N \cdot \frac{\partial}{\partial\Gamma_N} \tag{3.26}$$

The integration over Γ_N in the third equality is performed by parts. The first term vanishes due to the boundary condition at $\Gamma = \pm\infty$. We have used the obvious relation in the third equality, which signifies that the equilibrium distribution function does not depend on time, that is,

$$\hat{L}'f_N^{\mathrm{eq}}(\Gamma_N) = \dot{\Gamma}_N \cdot \frac{\partial f_N^{\mathrm{eq}}(\Gamma_N)}{\partial\Gamma_N} = 0 \tag{3.27}$$

Equation (3.24) is derived eventually as follows,

$$\begin{aligned}
\langle A\exp(t\hat{L}')B\rangle &= \sum_n \frac{t^n}{n!}\left\langle A\hat{L}'^n B\right\rangle = \sum_n \frac{t^n}{n!}(-1)^n\left\langle\left(\hat{L}'^n A\right)B\right\rangle \\
&= \langle(\exp(-t\hat{L}')A)B\rangle
\end{aligned} \tag{3.28}$$

From Eq. (3.28), one finds,

$$\langle\dot{B}(t)A\rangle = -\langle B\dot{A}(-t)\rangle = -\langle B(t)\dot{A}\rangle \tag{3.29}$$

and

$$\langle\dot{B}A\rangle = -\langle B\dot{A}\rangle \tag{3.30}$$

The relations are general, and valid for the case of $B = A$. Therefore,

$$\langle A\dot{A}\rangle = 0 \tag{3.31}$$

Equation (3.31) is telling that there is no correlation between a dynamic variable and its time derivative. In another word, a dynamic variable and its time derivative is "orthogonal" in the phase space.

Let us remark some an important property of the Liouville operator $\hat{L}'$ defined by Eq. (3.7). That is, the Liouville operator is "anti-Hermitian", meaning that the eigen value of the operator is *imaginary*. It is favorable to convert the operator into Hermitian for using good properties of the Hermitian operator, that is, all the eigen values of a Hermitian operator are *real*.

For the purpose, we define the new Liouville operator by

$$\hat{L} = -i\hat{L}' = i\sum_{i=1}^{3N}\left[\frac{\partial H}{\partial q_i}\frac{\partial}{\partial p_i} - \frac{\partial H}{\partial p_i}\frac{\partial}{\partial q_i}\right] \tag{3.32}$$

The Liouville equation of motion and the Liouville equation are rewritten in terms of the new operator as,

$$\frac{\mathrm{d}A}{\mathrm{d}t} = i\hat{L}A \quad \text{and} \quad \frac{\mathrm{d}f_N}{\mathrm{d}t} = -i\hat{L}f_N \tag{3.33}$$

As is shown in the following, the new operator is in fact Hermitian,

$$\begin{aligned}\langle -i\hat{L}AB^*\rangle &= -\langle A(-i\hat{L})^*B^*\rangle\\ -i\langle\hat{L}AB^*\rangle &= -i\langle A\hat{L}^*B^*\rangle\\ \langle\hat{L}AB^*\rangle &= \langle A(\hat{L}B)^*\rangle\end{aligned} \tag{3.34}$$

3.2 Generalized Langevin Equation

The generalized Langevin equation is an equation of motion to describe the time evolution of particular variable, referred to as *dynamic variables*, defined in the phase space. The appearance of the equation is similar to the phenomenological Langevin equation (PLE), having the frictional and random forces. However, there is a marked difference between the two equations. The difference is concerned with the origin of the frictional and random forces. In PLE, the forces are defined in terms of the phenomenological variables such as viscosity. On the other hand, those force are derived from the equation of motion in the phase space as a consequence of the projection of all degrees of freedom onto the dynamic variables. The section begins with an introduction of the generalized Langevin equation (GLE).

The time evolution of a variable A in the phase space can be described by the following equation (Liouville equation),

$$\frac{\mathrm{d}A(t)}{\mathrm{d}t} = i\hat{L}A(t) \tag{3.35}$$

In Eq. (3.35), $\hat{L}$ is the Liouville operator defined by,

$$\hat{L} \equiv i\sum_i \left[\left(\frac{\partial H}{\partial \mathbf{r}_i}\right) \cdot \left(\frac{\partial}{\partial \mathbf{p}_i}\right) - \left(\frac{\partial H}{\partial \mathbf{p}_i}\right) \cdot \left(\frac{\partial}{\partial \mathbf{r}_i}\right) \right]. \tag{3.36}$$

GLE is obtained by projecting variables B(t) onto a vector A, referred to as the dynamic variables. An operator ($\hat{P}$) to project variables onto the dynamic variable is defined by the following equation,

$$\hat{P}\mathbf{B}(t) = \frac{\mathbf{A} \cdot (\mathbf{A}, \mathbf{B}(t))}{(\mathbf{A}, \mathbf{A})} \tag{3.37}$$

or

$$\hat{P} \equiv \frac{\mathbf{A} \cdot (\mathbf{A}, \ )}{(\mathbf{A}, \mathbf{A})} \tag{3.38}$$

The projection operator satisfies the following relation, which can be proved by operating $\hat{P}$ onto the both sides of Eq. (3.37).

$$\hat{P}\hat{P} = \hat{P} \tag{3.39}$$

$B(t)$ in Eq. (3.37) is a variable defined in the phase space. (…, …) signifies the inner product of two vectors in the Hilbert space, defined by,

$$(\mathbf{A}, \mathbf{B}) = \langle \mathbf{A}\tilde{\mathbf{B}} \rangle = \int \mathrm{d}\Gamma f^{\mathrm{eq}}(\Gamma)\tilde{\mathbf{B}}(\Gamma), \tag{3.40}$$

where $f^{\mathrm{eq}}(\Gamma)$ represents a probability distribution function in equilibrium, and $\tilde{\mathbf{B}}$ means the adjoint vector of $\mathbf{B}$. The GLE can be obtained by applying the projection operator $\hat{P}$ to the Liouville Eq. (3.35), that is,

$$\frac{\mathrm{d}\mathbf{A}(t)}{\mathrm{d}t} = i\Omega\mathbf{A}(t) + \int_0^t \zeta(t-\tau)\mathbf{A}(\tau)\mathrm{d}\tau + \mathbf{f}(t) \tag{3.41}$$

in which $i\Omega$ is defined by,

$$i\Omega = \frac{(\dot{\mathbf{A}}, \mathbf{A})}{(\mathbf{A}, \mathbf{A})}. \tag{3.42}$$

The quantity $i\Omega$ defined by Eq. (3.42) is sometime referred as "Collective Frequency", because it is related to a *frequency* of the collective fluctuation of the dynamic variables. The physical significance of the quantity may become more transparent later when the theory is applied to the density fluctuation of water and the structural fluctuation of protein.

$\mathbf{f}(t)$ in Eq. (3.41) is a random force defined by

$$\mathbf{f}(t) \equiv \exp\left(i\hat{Q}\hat{L}t\right)\hat{Q}\dot{\mathbf{A}} \tag{3.43}$$

where $\hat{Q} \equiv 1 - \hat{P}$ is an operator to project the variables onto the space orthogonal to the space spanned by the dynamic variables $\mathbf{A}$. $\mathbf{f}(t)$ is uncorrelated with the dynamic variables as a consequence of the projection. That is why the quantity is referred as "random force".

$\zeta(t)$ in Eq. (3.41) is defined by

$$\zeta(t) = \frac{(\mathbf{f}(t), \mathbf{f}(0))}{(\mathbf{A}, \mathbf{A})} \tag{3.44}$$

The quantity is referred to as the "friction kernel", because it is a counter part of the friction coefficient in PLE. However, it is not just an instantaneous friction, but it has a past memory as is implied by the integral with respect to time in Eq. (3.41). In that reason, the friction kernel is referred to as *memory kernel*. The derivation of Eq. (3.44) is briefly reviewed as follows.

We start the derivation from the general solution of Eq. (3.35), that is,

$$\mathbf{A}(t) = \exp\left(i\hat{L}t\right)\mathbf{A}(0) = \exp\left(i\hat{L}t\right)\mathbf{A} \tag{3.45}$$

The derivative of Eq. (3.45) with respect to time t leads,

$$\frac{\mathrm{d}\mathbf{A}(t)}{\mathrm{d}t} = \exp\left(i\hat{L}t\right)i\hat{L}\mathbf{A} \tag{3.46}$$

By splitting the right-hand-side of Eq. (3.46) into two parts with the operators $\hat{P}$ and $\hat{Q}\left(\equiv 1 - \hat{P}\right)$, one finds,

$$\frac{\mathrm{d}\mathbf{A}(t)}{\mathrm{d}t} = \exp\left(i\hat{L}t\right)i\left(\hat{P} + \hat{Q}\right)\hat{L}\mathbf{A}. \tag{3.47}$$

The first term in the right-hand-side of Eq. (3.47) may be rewritten as

$$\exp\left(i\hat{L}t\right)i\hat{P}\hat{L}\mathbf{A} = i\exp\left(i\hat{L}t\right)\frac{\left(\hat{L}\mathbf{A}, \mathbf{A}\right)}{(\mathbf{A}, \mathbf{A})}\mathbf{A}$$

$$= \frac{(\dot{\mathbf{A}}, \mathbf{A})}{(\mathbf{A}, \mathbf{A})}\mathbf{A}(t) \equiv i\Omega\mathbf{A}(t) \tag{3.48}$$

where $i\Omega$ is defined by Eq. (3.42). This gives the first term in Eq. (3.47). The second term is manipulated by using the following identity,

$$\exp\left[\left(\hat{F} + \hat{G}\right)t\right] = \exp\left(\hat{F}t\right) + \int_0^t d\tau \exp\left[\left(\hat{F} + \hat{G}\right)(t - \tau)\right]\hat{G}\exp\left(\hat{F}\tau\right). \tag{3.49}$$

The equality is readily verified as follows. We multiply the both side by $\exp[-t\hat{F}]$ and then take the derivative of both sides with respect to time t. With the identifications, $\hat{F} = i\hat{Q}\hat{L}$ and $\hat{G} = i\hat{P}\hat{L}$,

$$\exp\left(i\hat{L}t\right) = \exp\left(i\hat{Q}\hat{L}t\right) + \int_0^t d\tau \exp\left[i\hat{L}(t - \tau)\right]i\hat{P}\hat{L}\exp\left(i\hat{Q}\hat{L}\tau\right) \tag{3.50}$$

We rewrite the second term in Eq. (3.47) as

$$\begin{aligned} \exp\left(i\hat{L}t\right)i\hat{Q}\hat{L}\mathbf{A} &= \exp\left[i\left(\hat{Q} + \hat{P}\right)\hat{L}t\right]\hat{Q}\dot{\mathbf{A}} \\ &= \exp\left[i\hat{Q}\hat{L}t\right]\hat{Q}\dot{\mathbf{A}} \\ &\quad + \int_0^t d\tau \exp\left[i\hat{Q}(t - \tau)\right]\left(i\hat{P}\hat{L}\right)\exp\left(i\hat{Q}\hat{L}\tau\right)\hat{Q}\dot{\mathbf{A}} \end{aligned} \tag{3.51}$$

The first term of the equation is the same as the random force term defined by Eq. (3.43), while the second term is identified as the friction term defined by Eq. (3.44) as follows,

$$\begin{aligned} \int_0^t d\tau \exp\left[i\hat{L}(t - \tau)\right]\left(\hat{P}i\hat{L}\right)\exp\left(i\hat{Q}\hat{L}\tau\right)\hat{Q}\dot{\mathbf{A}} &= \int_0^t d\tau \exp\left[i\hat{L}(t - \tau)\right]\left(\hat{P}i\hat{L}\right)\mathbf{f}(\tau) \\ &= \int_0^t d\tau \exp\left[i\hat{L}(t - \tau)\right]\frac{\left(i\hat{L}\mathbf{f}(\tau), \mathbf{A}\right)}{(\mathbf{A}, \mathbf{A})}\mathbf{A} \\ &= \int_0^t d\tau \exp\left[i\hat{L}(t - \tau)\right]\frac{\left(\mathbf{f}(\tau), i\hat{L}\mathbf{A}\right)}{(\mathbf{A}, \mathbf{A})}\mathbf{A} \\ &= \int_0^t d\tau \exp\left[i\hat{L}(t - \tau)\right]\frac{\left(\mathbf{f}(\tau), \dot{\mathbf{A}}\right)}{(\mathbf{A}, \mathbf{A})}\mathbf{A} \end{aligned}$$

$$= \int_0^t d\tau \frac{\left(\mathbf{f}(\tau), \dot{\mathbf{A}}\right)}{(\mathbf{A}, \mathbf{A})} \dot{\mathbf{A}}(t - \tau)$$

$$= \int_0^t d\tau \zeta(\tau) \dot{\mathbf{A}}(t - \tau) \tag{3.52}$$

where $\zeta(t)$ is the friction (or memory) kernel defined by Eq. (3.44). We have used the following obvious relations in the manipulation,

$$\left(\mathbf{f}(t), (1 - \hat{Q})\dot{\mathbf{A}}\right) = 0, \tag{3.53}$$

and

$$\begin{aligned} \left(\mathbf{f}(t), \dot{\mathbf{A}}\right) &= \left(\mathbf{f}(t), \hat{Q}\dot{\mathbf{A}}\right) \\ &= (\mathbf{f}(t), \mathbf{f}(0)) \end{aligned} \tag{3.54}$$

Thus, the derivation of GLE is completed.

GLE has many features that are common to PLE: the friction force and the random force are most typical ones. Most importantly, GLE inherits the fluctuation dissipation theorem that bridges between the random force $\mathbf{f}(t)$ and the friction $\zeta(t)$. Although the physical origin of "random force" and "friction" are the same, GLE has distinct features, which PLE does not have. GLE is a first principle theory, meaning that all the variables in the equation can be derived in principle from a Hamiltonian of the system. Two important factors in the equation should be calculated from the first principle: the collective frequency $i\Omega$ and the friction kernel $\zeta(t)$. As an example of an application of GLE to liquids, the structural dynamics of water is studied in the following section.

3.3 Generalized Langevin Equation of Water (Molecular Liquids)

In the preceding section, we have reviewed the generalized Langevin theory. In the present section, the theory is applied to the dynamics of *water*.

As been reviewed in Chap. 2, the spatial fluctuation of liquids is characterized by the density pair correlation function that is the variance of the spatial density fluctuation (Chap. 2). Therefore, the *dynamics* of liquids is equivalent to the time evolution of the *spatial density fluctuation and its variance* or *the density pair correlation function.* The temporal fluctuation of the density pair correlation function is referred to as the *space–time correlation function.* It is also called "van Hove function", honoring the person who has first introduced the function. Crucial for the theory to be successful is the choice of dynamic variables which are essential for describing the time evolution of the spatial density fluctuation.

The most important requirement for the dynamic variables to describe water is its capability to take account for the chemical characteristics represented by the *hydrogen-bonds*. The dynamics of water reflects the chemical specificity through the liquid structure, or the density pair correlation function. Then, how the chemical specificity of water will be incorporated in its dynamics?

A water molecule has a characteristic triangular-shape as everybody knows. The liquid structure features the hydrogen-bond between a pair of molecules, as been described briefly in Chap. 2. So, it is the question how those characteristics of water may be incorporated in its dynamics, or the space time correlation function.

There is a model to describe dynamics of water that has been practiced conventionally in most of the theories of liquid dynamics for long time. The model splits the dynamics into the translational and rotational motion. The model has been employed to interpret experimental data of relaxation in liquids, such as NMR (Abragam 1961). The model is simple, and its physics is easy to be understood. However, the theoretical description of dynamics in the liquid state becomes increasingly difficult. Such a theory employs the spherical-harmonics description for the rotational motion (upper picture in Fig. 3.1). Unfortunately, the convergence of the spherical-harmonic expansion become increasingly slow.

The entirely different model to describe the liquid dynamics has been introduced by the author in 1992 based on the interaction site model of liquids, which has been reviewed in Chap. 2 (Hirata 1992). According to the new model, the dynamics of a molecule is interpreted as a translational motion of each atom consisting the molecule, which is correlated each other. An image of the dynamics is illustrated in Fig. 3.1 (bottom) for a diatomic molecule as an example. When the two atoms (or

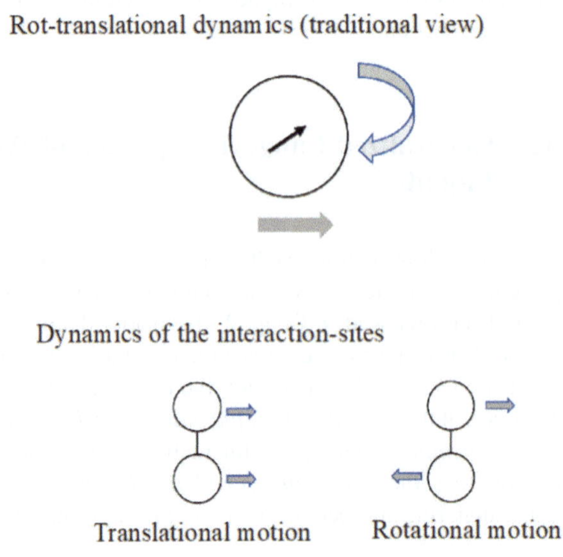

Fig. 3.1 Two models for the dynamics of a molecule in liquids

sites) in a molecule move in the *same* direction, the entire dynamics of the molecule is *translational* motion (Fig. 3.1, bottom left). On the other hand, the entire motion becomes *rotational*, when the two atoms move in the *opposite* direction (Fig. 3.1, bottom right). The actual dynamics of a molecule is composed of the two motions. It is the most important advantage of the model to make a bridge with the RISM theory that can describe the liquid structure of complex molecules such as tertiary butanol (Yoshida et al. 2002).

Generalized Langevin Equation of Molecular Liquids: Now, we formulate the generalized Langevin equation (GLE) of molecular liquids. The formulation is based on the interaction-site-model, described in Chap. 2. The Hamiltonian of the model is

$$H = \sum_{i=1}^{N}\sum_{a=1}^{n}\left[\frac{\mathbf{p}_i^a \cdot \mathbf{p}_j^a}{2m_a} + \sum_{j\neq i}\sum_{a\neq b} u\left(\left|\mathbf{r}_i^a - \mathbf{r}_j^b\right|\right)\right]. \tag{3.55}$$

In Eq. (3.55), $\mathbf{p}_i^a$ and m_a denote the momentum and the mass of atom a of the ith molecule, $u\left(\left|\mathbf{r}_i^a - \mathbf{r}_j^b\right|\right)$ is the interaction energy between atom a in the ith molecule and atom b in the jth molecule: N denotes the number of molecules in the system, and n the number of atoms in a molecule.

The Liouville operator (Eq. 3.36) is expressed by

$$i\hat{L} = \sum_{i=1}^{N}\sum_{a=1}^{n}\left[\frac{1}{2m_a}\mathbf{p}_i^a \cdot \frac{\partial}{\partial \mathbf{r}_i^a} - \sum_{j\neq i}\sum_{\beta\neq\alpha} \frac{\partial u\left(\left|\mathbf{r}_i^a - \mathbf{r}_j^b\right|\right)}{\partial \mathbf{r}_i^a} \cdot \frac{\partial}{\partial \mathbf{p}_i^a}\right] \tag{3.56}$$

The fluctuation of the number density ($\delta\rho^a$) of atoms and its conjugated momentum (J^a) are chosen as the dynamic variables,

$$\delta\rho^a(\mathbf{r}, t) = \sum_i \delta\rho\left(\mathbf{r} - \mathbf{r}_i^a(t)\right) - \rho_a \tag{3.57}$$

and

$$\mathbf{J}^a(t) = \sum_i \mathbf{p}_i^a \delta\left(\mathbf{r} - \mathbf{r}_i^a(t)\right) \tag{3.58}$$

We work on the Fourier space rather than the real space to derive GLE, because the derivation is much easier. In the Fourier space, the dynamic variable is expressed by,

$$\mathbf{A}(t) = \begin{pmatrix} \delta\rho_{\mathbf{k}}^a(t) \\ J_{\mathbf{k}}^a(t) \end{pmatrix} \tag{3.59}$$

where $\delta\rho_k^a$ and $\mathbf{J}_k^a$ are defined by

$$\delta\rho_{\mathbf{k}}^a(t) \equiv \int d\mathbf{r} \exp(i\mathbf{k}\cdot\mathbf{r})\delta\rho^a(\mathbf{r},t)$$
$$= \sum_i \exp\left[i\mathbf{k}\cdot\mathbf{r}_i^a(t)\right] - (2\pi)^3\rho_a\delta(\mathbf{k}) \tag{3.60}$$

and

$$\mathbf{J}_{\mathbf{k}}^a(t) = \int d\mathbf{r} \exp(i\mathbf{k}\cdot\mathbf{r}) \sum_i \mathbf{p}_i^a \delta\left(\mathbf{r}-\mathbf{r}_i^a(t)\right)$$
$$= \sum_i \mathbf{p}_i^a(t) \exp\left(i\mathbf{k}\cdot\mathbf{r}_i^a(t)\right) \tag{3.61}$$

In the following, we derive each term in the GLE (Eq. 3.41). The collective frequency term $i\Omega = \left(\dot{\mathbf{A}},\mathbf{A}\right)/(\mathbf{A},\mathbf{A})$ for molecules can be written in terms of the definition for $\delta\rho_k^a$ and $\mathbf{J}_k^a$ as

$$(\mathbf{A},\mathbf{A}) = \begin{pmatrix} \left(\delta\rho_{\mathbf{k}}^a, \delta\rho_{\mathbf{k}}^b\right) & \left(\mathbf{J}_{\mathbf{k}}^a, \delta\rho_{\mathbf{k}}^b\right) \\ \left(\delta\rho_{\mathbf{k}}^a, \mathbf{J}_{\mathbf{k}}^b\right) & \left(\mathbf{J}_{\mathbf{k}}^a, \mathbf{J}_{\mathbf{k}}^b\right) \end{pmatrix} \tag{3.62}$$

The off diagonal elements in the matrix are zero by definition of the dynamic variables, namely

$$\left(\delta\rho_{\mathbf{k}}^a, \mathbf{J}_{\mathbf{k}}^b\right) = 0, \quad \left(\mathbf{J}_{\mathbf{k}}^a, \delta\rho_{\mathbf{k}}^b\right) = 0 \tag{3.63}$$

We define the momentum correlation function by

$$J_{ab}(\mathbf{k}) \equiv \frac{1}{N}\left(\mathbf{J}_{-\mathbf{k}}^a, \mathbf{J}_{\mathbf{k}}^b\right) = \frac{1}{N}\left\langle \mathbf{J}_{-\mathbf{k}}^a \cdot \mathbf{J}_{\mathbf{k}}^b \right\rangle \tag{3.64}$$

$$J_{ab}(\mathbf{k}) = \frac{1}{N}\sum_i\sum_j \frac{1}{m_a}\frac{1}{m_b}\left\langle \mathbf{p}_i^a \cdot \mathbf{p}_j^b \exp\left[-i\mathbf{k}\cdot\left(\mathbf{r}_i^a - \mathbf{r}_j^b\right)\right]\right\rangle$$
$$= \frac{1}{N}\sum_i\sum_j \frac{1}{m_a}\frac{1}{m_b}\left\langle \mathbf{p}_i^a \cdot \mathbf{p}_j^b\right\rangle\left\langle \exp\left[-i\mathbf{k}\cdot\left(\mathbf{r}_i^a - \mathbf{r}_j^b\right)\right]\right\rangle$$
$$= \frac{1}{N}\sum_i\sum_j \left\langle \mathbf{v}_i^a \cdot \mathbf{v}_j^b\right\rangle \delta_{ij}\left\langle \exp\left[-i\mathbf{k}\cdot\left(\mathbf{r}_i^a - \mathbf{r}_j^b\right)\right]\right\rangle$$
$$= \frac{1}{N}\sum_i \left\langle \mathbf{v}_i^a \cdot \mathbf{v}_j^b\right\rangle\left\langle \exp\left[-i\mathbf{k}\cdot\left(\mathbf{r}_i^a - \mathbf{r}_j^b\right)\right]\right\rangle \tag{3.65}$$

where we have used the obvious relation $\left\langle \mathbf{v}_i^a \cdot \mathbf{v}_j^b\right\rangle = \left\langle \mathbf{v}_i^a\right\rangle\left\langle \mathbf{v}_j^b\right\rangle = 0$ if $i \neq j$. Then, the correlation matrix $(\mathbf{A},\mathbf{A})$ and its inverse become,

$$(\mathbf{A}, \mathbf{A}) = \begin{pmatrix} N\chi_{ab}(\mathbf{k}) & 0 \\ 0 & J_{ab}(\mathbf{k}) \end{pmatrix} \tag{3.66}$$

$$(\mathbf{A}, \mathbf{A})^{-1} = \begin{pmatrix} N^{-1}\chi_{ab}(\mathbf{k})^{-1} & 0 \\ 0 & J_{ab}(\mathbf{k})^{-1} \end{pmatrix} \tag{3.67}$$

where the matrix inverse is defined by

$$\sum_{c=1}^{n} \chi_{ac}(\mathbf{k})\chi_{cb}(\mathbf{k})^{-1} = \delta_{ab}, \quad \sum_{c=1}^{n} J_{ac}(\mathbf{k})J_{cb}(\mathbf{k})^{-1} = \delta_{ab} \tag{3.68}$$

The correlation matrix $(\mathbf{A}, \dot{\mathbf{A}})$ reads,

$$\left(\mathbf{A}, \dot{\mathbf{A}}\right) = \begin{pmatrix} \left(\delta\rho_{\mathbf{k}}^{a}, \dot{\rho}_{\mathbf{k}}^{b}\right) & \left(\mathbf{J}_{\mathbf{k}}^{a}, \dot{\rho}_{\mathbf{k}}^{b}\right) \\ \left(\delta\rho_{\mathbf{k}}^{a}, \dot{\mathbf{J}}_{\mathbf{k}}^{b}\right) & \left(\mathbf{J}_{\mathbf{k}}^{a}, \dot{\mathbf{J}}_{\mathbf{k}}^{b}\right) \end{pmatrix} \tag{3.69}$$

where $\dot{\rho}_{\mathbf{k}}^{a} = i\hat{L}\rho_{\mathbf{k}}^{a}$, and $\dot{\mathbf{J}}_{\mathbf{k}}^{a}$ is

$$\dot{\mathbf{J}}_{\mathbf{k}}^{a} = \sum_{i} \frac{1}{m_a}\left(\dot{\mathbf{p}}_i^a + \mathbf{p}_i^a i\mathbf{k} \cdot \frac{\mathbf{p}_i^a}{m_a}\right) \exp\left(i\mathbf{k} \cdot \mathbf{r}_i^a\right). \tag{3.70}$$

Considering that $\left(\delta\rho_{\mathbf{k}}^{a}, \dot{\rho}_{\mathbf{k}}^{b}\right) = 0$, one finds the following expression for the collective frequency term,

$$i\Omega = \frac{(\mathbf{A}, \dot{\mathbf{A}})}{(\mathbf{A}, \mathbf{A})} = \begin{pmatrix} 0 & ik\delta_{ab} \\ i\mathbf{k}\sum_{c=1}^{n} J_{ac}(\mathbf{k})\chi_{cb}(\mathbf{k})^{-1} & 0 \end{pmatrix} \tag{3.71}$$

Then, the *steady-state* part, the collective frequency, of the generalized Langevin equation is given by,

$$i\Omega \cdot \mathbf{A}(t) = \begin{pmatrix} i\mathbf{k} \cdot \mathbf{J}_{k}^{a}(t) \\ ik\sum_{c}^{n}\sum_{b}^{n} J_{ac}(\mathbf{k})\chi_{cb}(\mathbf{k})^{-1}\delta\rho_{\mathbf{k}}^{b}(t) \end{pmatrix} \tag{3.72}$$

The Fluctuating Force (Random Force): From Eq. (3.43), the fluctuating force at $t = 0$ is given by

$$\mathbf{f} = \left(1 - \hat{P}\right)\dot{\mathbf{A}} = \dot{\mathbf{A}} - i\Omega \cdot \mathbf{A} \tag{3.73}$$

where $\hat{P}\dot{\mathbf{A}} = (\mathbf{A}, \dot{\mathbf{A}}) \cdot (\mathbf{A}, \mathbf{A})^{-1}\mathbf{A} = i\Omega \cdot \mathbf{A}$ is taken into account. $\dot{\mathbf{A}}$ and $i\Omega \cdot \mathbf{A}$ are

$$\dot{\mathbf{A}} = \begin{pmatrix} \dot{\rho}_{\mathbf{k}}^{a} \\ \dot{\mathbf{j}}_{\mathbf{k}}^{a} \end{pmatrix} = \begin{pmatrix} i\mathbf{k} \cdot \mathbf{J}_{\mathbf{k}}^{a} \\ \dot{\mathbf{j}}_{\mathbf{k}}^{a} \end{pmatrix} \tag{3.74}$$

$$i\Omega \cdot \mathbf{A}(t) = \begin{pmatrix} i\mathbf{k} \cdot \mathbf{J}_{k}^{a}(t) \\ ik \sum_{c}^{n} \sum_{b}^{n} J_{ac}(\mathbf{k}) \chi_{cb}(\mathbf{k})^{-1} \delta\rho_{\mathbf{k}}^{b}(t) \end{pmatrix} \tag{3.75}$$

Then, the fluctuating force at $t = 0$ becomes

$$\mathbf{f} = \begin{pmatrix} 0 \\ \Xi_{\mathbf{k}}^{a} \end{pmatrix} \tag{3.76}$$

where $\Xi_{\mathbf{k}}^{a}$ is defined by

$$\Xi_{\mathbf{k}}^{a} = \dot{\mathbf{j}}_{\mathbf{k}}^{a} - i\mathbf{k} \sum_{b}^{n} \sum_{c}^{n} J_{ac}(\mathbf{k}) \chi_{cb}(\mathbf{k})^{-1} \delta\rho_{\mathbf{k}}^{b}(t) \tag{3.77}$$

The fluctuating force is derived as

$$\mathbf{f}(t) = \exp\left[it(1 - \hat{P})\hat{L}\right]\mathbf{f} = \exp\left[it(1 - \hat{P})\hat{L}\right] \begin{pmatrix} 0 \\ \Xi_{\mathbf{k}}^{a} \end{pmatrix} \tag{3.78}$$

The Memory Kernel (Friction Matrix): Using the definition in terms of the fluctuating force, we get the following expression for the memory kernel,

$$\begin{aligned} \mathbf{K}(t) &\equiv \frac{(\mathbf{f}, \mathbf{f}(t))}{(\mathbf{A}, \mathbf{A})} \\ &= \begin{pmatrix} \mathbf{0} & \mathbf{0} \\ \mathbf{0} & N^{-1} J_{ac}^{-1} \left(\Xi_{\mathbf{k}}^{c}, \Xi_{\mathbf{k}}^{b}(t) \right) \end{pmatrix} \end{aligned} \tag{3.79}$$

where $\Xi_{\mathbf{k}}^{a}(t) \equiv \exp\left(it(1 - \hat{P})\hat{L}\right) \Xi_{\mathbf{k}}^{a}$.

Generalized Langevin Equations of Molecular Liquids: Summarizing all the terms derived in Eqs. (3.75), (3.78), (3.79) into Eq. (3.41), the following equations are obtained.

$$\frac{\mathrm{d}\delta\rho_{\mathbf{k}}^{a}(t)}{\mathrm{d}t} = i\mathbf{k} \cdot \mathbf{J}_{\mathbf{k}}^{a}(t) \tag{3.80}$$

$$\frac{\mathrm{d}\mathbf{J}_{k}^{a}(t)}{\mathrm{d}t} = ik \sum_{b}^{n} \sum_{c}^{n} J_{ac}(\mathbf{k}) \chi_{cb}(\mathbf{k})^{-1} \delta\rho_{\mathbf{k}}^{b}(t)$$

$$-\sum_{b}^{n}\sum_{c}^{n}\int_{0}^{t} \mathrm{d}s M_{\mathbf{k}}^{bc}(t-s) + \Xi_{\mathbf{k}}^{a}(t) \tag{3.81}$$

where

$$M_{\mathbf{k}}^{bc}(s) = N^{-1}J_{ac}^{-1}(k)\left(\Xi_{\mathbf{k}}^{c}, \Xi_{\mathbf{k}}^{b}(s)\right). \tag{3.82}$$

The first and second equations correspond, respectively, to the continuity equation and the equation of motion (or the momentum conservation) in the hydrodynamics. By taking the derivative of the first equation with respect to time t, and by substituting the second equation into $\dot{\mathbf{J}}_{k}^{a}(t)$, one finds the GLE for the density fluctuation as follows.

$$\frac{\mathrm{d}^2\delta\rho_{\mathbf{k}}^{a}(t)}{\mathrm{d}t^2} = i\mathbf{k}\cdot\left[i\mathbf{k}\sum_{b}^{n}\sum_{c}^{n} J_{ac}\chi_{cb}(\mathbf{k})^{-1}\delta\rho_{\mathbf{k}}^{b}(t) - \sum_{b}^{n}\sum_{c}^{n}\int_{0}^{t} \mathrm{d}s\zeta_{\mathbf{k}}^{bc}(t-s) + \Xi_{\mathbf{k}}^{a}(t)\right]. \tag{3.83}$$

The expression can be Fourier transformed back to the real space to find,

$$\frac{\partial^2\delta\rho_a(\mathbf{r},t)}{\partial t^2} = -\sum_{b,c}\int \kappa_{ab}^{\mathrm{solv}}(\mathbf{r},\mathbf{r}')\delta\rho_c(\mathbf{r}',t)\mathrm{d}\mathbf{r}' + \left(\frac{1}{N}\sum_{b,c}\int \mathrm{d}\mathbf{r}'\int \mathrm{d}\mathbf{r}'' J_{ab}(\mathbf{r},\mathbf{r}')\int_{0}^{t} \mathrm{d}s\zeta_{bc}(\mathbf{r}';t-s)\cdot\frac{\mathrm{d}\delta\rho_c(\mathbf{r}'',s)}{\mathrm{d}t}\mathrm{d}s\right) - \Xi_a(\mathbf{r},t) \tag{3.84}$$

where $\kappa_{ab}^{\mathrm{solv}}(\mathbf{r},\mathbf{r}')$ is defined as,

$$\kappa_{ab}^{\mathrm{solv}}(\mathbf{r},\mathbf{r}') \equiv -\sum_{c}\int \mathrm{d}\mathbf{r}'' J_{ac}(\mathbf{r},\mathbf{r}'')\chi_{cb}^{(2)}(\mathbf{r}'',\mathbf{r}')^{-1}. \tag{3.85}$$

Equation (3.84) embodies an extremely important physics concerning the density fluctuation in water. The equation implies that *the force* to restore the equilibrium density from a fluctuated state is *linear* to the *density fluctuation* itself. It means that the temporal fluctuation of the density around a biomolecule is *harmonic*, and that the probability distribution of the density fluctuation is *Gaussian*. Further, $\kappa_{ab}^{\mathrm{solv}}(\mathbf{r},\mathbf{r}')$ defined by Eq. (3.85) has a meaning of *Hessian* or the *force constant* of the restoring force.

3.4 The Density Fluctuation in Water: Space–Time Correlation (Van Hove) Function

An equation for the space–time correlation function, which describes the temporal relaxation of the density fluctuation in water, can be derived from Eq. (3.84) as follows.

Multiplying the both sides of the equation by $\delta\rho^b_{-\mathbf{k}}(0)$, and taking the ensemble average, one gets,

$$\frac{\mathrm{d}^2\langle\delta\rho^a_{\mathbf{k}}(t)\delta\rho^b_{-\mathbf{k}}(0)\rangle}{\mathrm{d}t^2} = -\sum_c \langle\omega^2_{\mathbf{k}}\rangle_{ac}\langle\delta\rho^a_{\mathbf{k}}(t)\delta\rho^b_{-\mathbf{k}}(0)\rangle - \sum_c \int_0^t \mathrm{d}s\zeta^{bc}_{\mathbf{k}}(t-s)\langle i\mathbf{k}\cdot\mathbf{J}^{bc}_{\mathbf{k}}(t)\delta\rho^b_{-\mathbf{k}}(0)\rangle \tag{3.86}$$

where the orthogonality between the fluctuating force and the dynamic variable, $\langle\Xi^a_{\mathbf{k}}(t)\delta\rho^b_{\mathbf{k}}(0)*\rangle = 0$, was considered, and the *corrective frequency* $\langle\omega^2_k\rangle_{ab}$ is defined by

$$\langle\omega^2_{\mathbf{k}}\rangle_{ac} = k^2\sum_c J_{ac}(k)\left(\chi(k)^{-1}\right)_{cb} \tag{3.87}$$

With the definition $F^{ab}_{\mathbf{k}}(t) \equiv N^{-1}\langle\delta\rho^a_{\mathbf{k}}(t)\delta\rho^b_{-\mathbf{k}}(0)\rangle$, one gets the following equation in the matrix notation,

$$\frac{\mathrm{d}^2\mathbf{F}_{\mathbf{k}}(t)}{\mathrm{d}t^2} = -\omega^2_{\mathbf{k}}\mathbf{F}_{\mathbf{k}}(t) - \int_0^t \mathrm{d}s\zeta_{\mathbf{k}}(t-s)\cdot\dot{\mathbf{F}}_{\mathbf{k}}(s) \tag{3.88}$$

where $\dot{\mathbf{F}}_{\mathbf{k}}(s) \equiv \mathrm{d}\mathbf{F}_{\mathbf{k}}(s)/\mathrm{d}s$ and $\delta\dot{\rho}^\alpha_{\mathbf{k}}(t) = i\mathbf{k}\cdot\mathbf{J}^\alpha_{\mathbf{k}}(t)$ are taken into account.

Diffusion Limit: Site–Site Smoluchowski–Vlasov (SSSV) Equation

As an instructive purpose, let us solve Eq. (3.88) for water in the diffusion limit, where the friction loses the past memory (or Markov process), and inertia term can be ignored, that is,

$$\zeta_{\mathbf{k}}(t-s) \sim \zeta_{\mathbf{k}}\delta(t-s) \tag{3.89}$$

$$\frac{\mathrm{d}^2\mathbf{F}_{\mathbf{k}}(t)}{\mathrm{d}t^2} = 0 \tag{3.90}$$

Then, the equation simplified considerably to read,

$$\frac{d}{dt}\mathbf{F}_{\mathbf{k}}(t) = -k^2\mathbf{J}(\mathbf{k})\chi(\mathbf{k})^{-1}\zeta_k^{-1}\mathbf{F}_k(t) \tag{3.91}$$

Defining the diffusion constant by $\mathbf{D}_{\mathbf{k}} \equiv \mathbf{J}(\mathbf{k})\zeta_{\mathbf{k}}^{-1}$, one gets a simple diffusion equation for a space–time correlation function of molecular liquids as,

$$\frac{d}{dt}\mathbf{F}_{\mathbf{k}}(t) = -k^2\mathbf{D}_k\chi(k)^{-1}\mathbf{F}_{\mathbf{k}}(t) \tag{3.92}$$

The equation referred to as "Site–Site Smoluchowski–Vlasov (SSSV)" equation has been derived by the author in 1992 (Hirata 1992).

Shown in Fig. 3.2 is the space–time correlation function (solid lines, dashed lines, and dotted lines) calculated by the SSSV theory, compared with the results from a molecular dynamics simulation (o, +, *). It is seen that the results from SSSV reproduce the simulation ("experimental") results fairly well (solid lines, $t = 0.01$ ps; dashed lines, $t = 0.1$ ps; dotted lines, 1.0 ps).

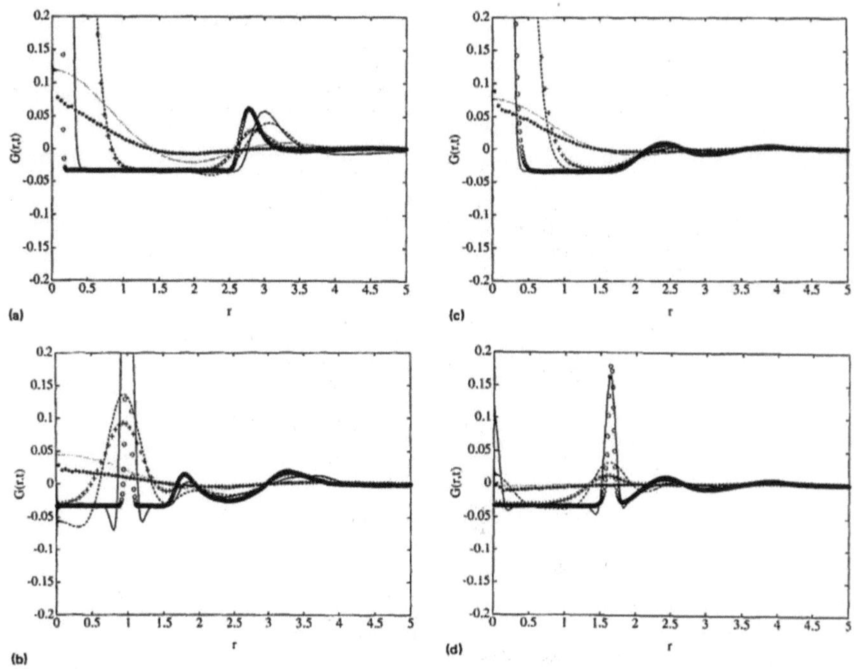

Fig. 3.2 Space–time correlation function of water calculated by the SSSV theory

3.5 Collective Density Fluctuations in Water

Figure 3.2 in the previous section indicates that both the spatial and temporal density fluctuations decay with time due to the friction. Then, it may raise a question. Is there any relation between the spatial and temporal density fluctuations? The section concerns this question.

The question is rephrased as follows, "Is there any relation between the *wavenumber* (an inverse of wave-length) and *frequency* of the density fluctuation?", because the wavenumber is the measure of the spatial fluctuation, while the frequency is that of temporal fluctuations.

As an example, let's imagine the sound propagation in water. A sound propagates through a medium with a density fluctuation. The density of the media will show a sigmoidal change in space as well as in time. The distance between the two tops of the wave *in space* is defined as the wave length λ, and its inverse as the wavenumber, $k = 2\pi/\lambda$. The time difference between the two tops of the wave *in time* is called "period", T, and its inverse as "frequency", ω. If one can observe the molecular motion of sound wave in one phase, many molecules will be in a *translational motion*, moving *cooperatively* in the same direction. It may be intuitively understood that longer the wave length λ, longer the period T: in another words, smaller the wavenumber k, less the frequency ω. Such a relation between the wavenumber and the frequency is referred to as *dispersion relation.* Let us explore the dispersion relation in water based on the generalized Langevin equation derived in the previous section (Chong and Hirata 1999a, b).

For that purpose, let's go back to Eq. (3.88), describing the time evolution of density correlation functions. If one ignores the friction term, he finds the equation describing the *steady-state* density-fluctuation in time,

$$\frac{d^2\delta\mathbf{F}_{\mathbf{k}}(t)}{dt^2} = -\langle\omega_{\mathbf{k}}^2\rangle\mathbf{F}_{\mathbf{k}}(t) \tag{3.93}$$

where $\langle\omega_k^2\rangle$ is the collective frequency matrix defined by Eq. (3.87), or

$$\langle\omega_{\mathbf{k}}^2\rangle_{ab} = k^2\sum_c J_{ac}(k)\left(\chi(k)^{-1}\right)_{cb} \tag{3.94}$$

Equation (3.93) is *formally* identical to that describing a *harmonic oscillator*, if one identifies $\langle\omega_k^2\rangle$ as a *force constant*. The equation implies that *the restoring force exerted on density fluctuation is proportional to the density fluctuation*, and that the proportional constant (or a "force constant") is $\langle\omega_k^2\rangle$. It is as if the same law as "Hook's law" in the harmonic oscillator is governing the density fluctuation. Then, with an analogy to the normal mode analysis, "*normal modes*" of the density fluctuation of atoms corresponding to "*eigen frequencies*" can be obtained by diagonalizing the matrix $\langle\omega_k^2\rangle$.

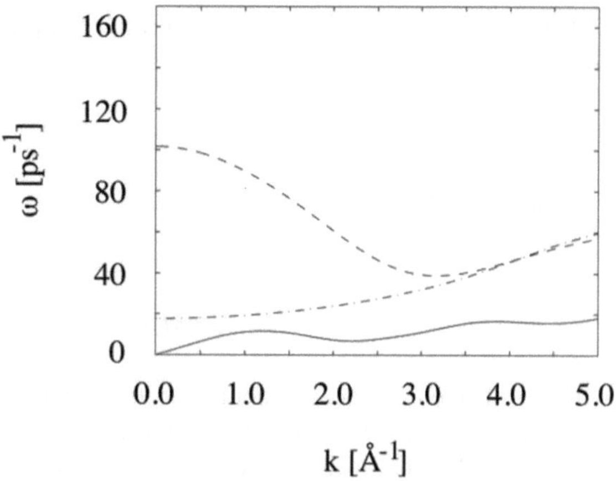

Fig. 3.3 The dispersion relation of the density fluctuation in water

The diagonalization corresponds to a transformation of the dynamic variables from the density of atoms to their linear combination,

$$\delta\rho_i(k) = \sum_a x_{ia}(k)\delta\rho_a(k) \tag{3.95}$$

where $\delta\rho_i(k)$ denotes the ith mode of the collective density fluctuation, $\delta\rho_a(k)$ denotes the density fluctuation of atom a. In the case of water, a distinguishes an oxygen atom (O) and two hydrogen atoms (H_1, H_2). Since water has three atoms, we have three eigen modes.

The eigen values (frequency) and eigen functions (modes of atomic density fluctuation) of $\langle\omega_k^2\rangle$ for water are depicted against the wavenumber in Fig. 3.3.

Now, let us clarify the physics of each mode. One of the three eigen frequencies exhibits the following asymptotic behavior at the limit of $k = 0$,

$$\omega_{\text{acou}}^2(k \to 0) = \frac{k_B T}{M\chi(0)}k^2 \tag{3.96}$$

where M denotes the mass of a molecule. The equation shows that the frequency (ω) at $k \to 0$ is proportional to the wavenumber $k = 1/\lambda$ (λ; wave length), that is,

$$\omega = ck, \tag{3.97}$$

where c is defined as

$$c = \sqrt{\frac{k_B T}{M\chi(0)}}. \tag{3.98}$$

Here, $\chi(0)$ is related to the isothermal compressibility χ_T by the following equation,

$$\rho k_B T \chi_T = \chi(0) \tag{3.99}$$

From Eqs. (3.98) and (3.99), one finds

$$c = \sqrt{\frac{1}{\rho M \chi_T}} = \sqrt{\frac{1}{\rho_M \chi_T}} \tag{3.100}$$

where ρ_M is the mass density defined from the number density as $\rho_M \equiv \rho \times M$.

The expression suggests that the proportional constant c in Eq. (3.97) is essentially a *sound* velocity, because it is the well-regarded hydrodynamic-relation between the compressibility and the sound velocity. However, there is a subtle difference from the hydrodynamic expression. The compressibility included in the hydrodynamic expression is *adiabatic*, while the compressibility here is *isothermal*. Nonetheless, the essential physics involved in the two expressions is the same: the sound velocity in a less compressible liquid is faster than that in a more compressible one. The mode satisfies the dispersion relation of the sound wave, $c = \lambda\omega$, thereby it is assigned to an acoustic mode. It is also understood that the mode is related to the *translational* motion of the molecule, since Eq. (3.100) includes the mass of a molecule, M.

On the other hand, the dispersion relations of the other two modes behave differently from the acoustic mode: the frequency does not converge to zero as the wavenumber becomes zero. The authors found that the fluctuational modes corresponding to the two curves assigned to the *rotational fluctuation* of water molecules around the two principal axes (Fig. 3.4). Such a rotational fluctuation of molecules induces the electric polarization, since water molecules have a dipole moment, and the polarization responds to the electromagnetic field of *light*. In that reason, such a mode of fluctuation is referred to as "optical modes". As can be readily understand from the experiment of the dielectric relaxation, the polarization can extend over *space* in macroscopic scale between two electrodes. On the other hand, the temporal fluctuation of the polarization is induced by rotational motions of water molecules, the time scale of which is rather short such as ~ 10^{-10} to ~ 10^{-11} s. It is the reason why the frequency of the optical modes does not converge to zero.

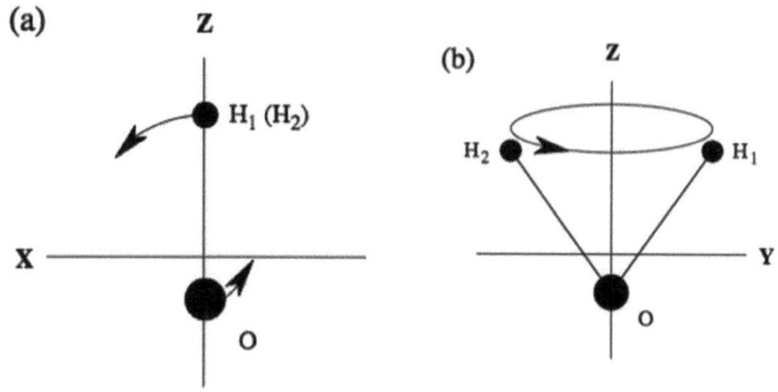

Fig. 3.4 Two rotational modes of a water molecule

References

Abragam A (1961) The principle of the nuclear magnetism. Clarendon Press, Oxford

Chong S-H, Hirata F (1999a) Interaction-site-model description of collective excitations in liquid water. I: theoretical study. J Chem Phys 111(7):3083–3094

Chong S-H, Hirata F (1999b) Interaction-site-model description of collective excitations in liquid water. II: comparison with simulation results. J Chem Phys 111(7):3095–3104

Friedman HL (1985) A course in statistical mechanics. Prentice-Hall, New Jersey

Hirata F (1992) Interaction-site representation of the Smoluchowski–Vlasov equation: the space-time correlation functions in a molecular liquid. J Chem Phys 96:4619–4624

Mori H (1958) Statistical-mechanical theory in transport in fluids. Phys Rev 112:1829

Mori H (1962) Collective motion of particles at finite temperatures. Prog Theor Phys (Jpn) 27:427

Mori H (1965) Transport, collective motion and Brownian motion. Prog Theor Phys (Jpn) 33:423–455

Yoshida K, Kovalenko A, Yamaguchi T, Hirata F (2002) Structure of tert-butyl alcohol-water mixtures studied by the RISM theory. J Phys Chem B 106(19):5042

Chapter 4
Generalized Langevin Theory of a Biomolecule-in-Water System

In Chap. 3, we have derived a generalized Langevin equation for a system of pure water, that describes the temporal as well as spatial density fluctuation of in water [(Eq. 3.84) in Chap. 3]. The most important finding there is that the temporal density fluctuation is *harmonic*, or *Gaussian*. The Gaussian character has a great advantage for a theoretical treatment, since it allows a *harmonic analysis* of the fluctuation.

In this chapter, we derive generalized Langevin equations (GLE) for the system of a biomolecule-*in*-water. The structural fluctuation of a biomolecule is subject to the density fluctuation of water, while the density fluctuation of water around a biomolecule is under strong influence of the biomolecule. Therefore, the generalized Langevin equation should account for the two types of fluctuation in a conjugated manner (Kim and Hirata 2013).

Our main concern here is a system of a biomolecule immersed in water at infinite dilutions. So, in the formulation, we consider a general solute–solvent system. In particular, we consider a solute molecule consisting of N_u atoms immersed in solvent consisting of N molecules, each having n atoms.

In the following section, we derive essentially two GLEs, one for the density fluctuation of water interacting with a biomolecule, and the other for the structural fluctuation of a biomolecule immersed in water. In Sect. 4.2, we analyze the physics involved in GLE for the structural fluctuation of a biomolecule in water, derived in Sect. 4.1. Section 4.3 is devoted to analyze the physics of density fluctuation of water around a biomolecule. As a consequence of the analyses, we conclude that the probability distributions of the both fluctuations are strictly *Gaussian*. In Sect. 4.4, we derive formula which relate the density and structural fluctuations to respective entropies.

F. Hirata, *Molecular Theory of Fluctuation in Life Phenomena*,
https://doi.org/10.1007/978-981-96-3364-7_4

4.1 Generalized Langevin Equations for a Solute–Solvent System

The Hamiltonian of the solute–solvent system is given by

$$H = H_0 + H_1 + H_2 \tag{4.1}$$

$$H_0 = \sum_{i=1}^{N}\sum_{a=1}^{n}\left[\frac{\mathbf{p}_i^a \cdot \mathbf{p}_i^a}{2m_a} + \sum_{j\neq i}\sum_{b\neq a} U_0\left(\left|\mathbf{r}_i^a - \mathbf{r}_j^b\right|\right)\right] \tag{4.1.1}$$

$$H_1 = \sum_{\alpha=1}^{N_u}\left[\frac{\mathbf{P}_\alpha \cdot \mathbf{P}_\beta}{2M_\alpha} + \sum_{\beta\neq\alpha} U_1\left(\left|\mathbf{R}_\alpha - \mathbf{R}_\beta\right|\right)\right] \tag{4.1.2}$$

$$H_2 = \sum_{\alpha=1}^{N_u}\sum_{i=1}^{N}\sum_{a=1}^{n} U_{\text{int}}\left(\left|\mathbf{R}_\alpha - \mathbf{r}_i^a\right|\right) \tag{4.1.3}$$

where M_α denotes the mass of the αth atom in the solute molecule, and m_a the mass of ath atom in a solvent molecule. The Hamiltonian of the solvent is denoted by H_0 where $\mathbf{r}_i^a$ and $\mathbf{p}_i^a$ are respectively the position and momentum of ath atom in the ith molecule of the solvent, and $U_0\left(r_{ij}^{ab}\right)$ $\left(r_{ij}^{ab} \equiv \left|\mathbf{r}_i^a - \mathbf{r}_j^b\right|\right)$ is the pair potential energy between them. H_1 is the Hamiltonian of the N_u solute atoms, and $\mathbf{R}_\alpha$ and $\mathbf{P}_\alpha$ are the position and momentum of the αth solute atoms (we use the Greek index for denoting the solute atoms), and $U_{\text{int}}\left(\left|\mathbf{R}_\alpha - \mathbf{r}_i^a\right|\right)$ is the interaction potential energy between the αth solute atom and the ath atom of the ith molecule in the solvent.

The associated Liouville operator $i\hat{L}$ is defined by

$$i\hat{L} = i\hat{L}_0 + i\hat{L}_1 \tag{4.2}$$

$$\begin{aligned} i\hat{L}_0 \equiv \sum_{i=1}^{N}\sum_{a=1}^{n}\Bigg[&\frac{1}{m_a}\mathbf{p}_i^a \cdot \frac{\partial}{\partial \mathbf{r}_i^a} \\ &-\sum_{j\neq i}\sum_{b\neq a}\frac{\partial U_0\left(\mathbf{r}_{ij}^{ab}\right)}{\partial \mathbf{r}_i^a}\cdot\frac{\partial}{\partial \mathbf{p}_i^a} - \sum_{\alpha=1}^{N_\text{u}}\frac{\partial U_{\text{int}}\left(\left|\mathbf{R}_\alpha - \mathbf{r}_i^a\right|\right)}{\partial \mathbf{r}_i^a}\cdot\frac{\partial}{\partial \mathbf{p}_i^a}\Bigg] \end{aligned} \tag{4.2.1}$$

$$i\hat{L}_1 = \sum_{\alpha=1}^{N_\text{u}}\left[\frac{\mathbf{P}_\alpha}{M_\alpha}\cdot\frac{\partial}{\partial \mathbf{R}_\alpha} + \mathbf{F}_\alpha \cdot \frac{\partial}{\partial \mathbf{P}_\alpha}\right] \tag{4.2.2}$$

where $\mathbf{F}_\alpha \equiv \mathbf{F}_\alpha^{(u)} + \mathbf{F}_\alpha^{(v)}$, and $\mathbf{F}_\alpha^{(u)}$ is the force exerted on the αth solute atom by the other solute atoms, $\mathbf{F}_\alpha^{(v)}$ the force exerted on the same solute atom by the solvent molecules. They are expressed explicitly by,

$$\mathbf{F}_\alpha^{(u)} = -\sum_{\beta \neq \alpha} \frac{\partial U_1 \left(\mathbf{R}_{\alpha\beta} \right)}{\mathbf{R}_\alpha} \tag{4.3}$$

$$\mathbf{F}_\alpha^{(v)} = -\sum_{i=1}^{N} \sum_{a=1}^{n} \frac{\partial U_{\mathrm{int}}\left(\left| \mathbf{R}_\alpha - \mathbf{r}_i^a \right| \right)}{\partial \mathbf{R}_\alpha} \tag{4.4}$$

It is the crucial step in the generalized Langevin theory to define the dynamic variables onto which all other variables in the phase space are projected. Why is it essential for the theory being successful? There are several reasons for that. The first is that the variables should be chosen appropriately in order for the theory to be able to describe the essential physics of processes under concern. The physics concerned with the present chapter is the *structural fluctuation* of protein and the *density fluctuation* of water, which are conjugated with each other. The second reason is that the variables should be readily handled by the statistical mechanics. Such considerations let us choose the following vector in the phase space as the dynamic variable $\mathbf{A}(t)$.

$$\mathbf{A}(t) = \begin{pmatrix} \Delta \mathbf{R}_\alpha(t) \\ \mathbf{P}_\alpha(t) \\ \delta\rho_{\mathbf{k}}^a(t) \\ \mathbf{J}_{\mathbf{k}}^a(t) \end{pmatrix} \tag{4.5}$$

In the equation, $\Delta \mathbf{R}_\alpha(t)$ represents the displacement of the αth atom of solute from its equilibrium position, defined by

$$\Delta \mathbf{R}_\alpha(t) = \mathbf{R}_\alpha(t) - \langle \mathbf{R}_\alpha \rangle, \tag{4.6}$$

where $\mathbf{R}_\alpha(t)$ denotes the position of atom α at time t, and $\langle \mathbf{R}_\alpha \rangle$ represents its equilibrium position. Therefore, it represents the structural fluctuation of a solute molecule. $\mathbf{P}_\alpha(t)$ is defined by $\mathbf{P}_\alpha(t) = M_\alpha \Delta \dot{\mathbf{R}}_\alpha(t) = M_\alpha \dot{\mathbf{R}}_\alpha(t)$, and $\delta\rho_{\mathbf{k}}^a(t)$ is the Fourier transform of the density fluctuation $\delta\rho^a(\mathbf{r}, t) \equiv \rho^a(\mathbf{r}, t) - \rho_0^a$ of solvent, in which $\rho^a(\mathbf{r}, t)$ is the density field of the ath atom of the solvent molecule, and ρ_0^a is the average number density.

$$\delta\rho^a(\mathbf{r}, t) \equiv \sum_i \delta\left(\mathbf{r} - \mathbf{r}_i^a(t) \right) - \rho_0^a \tag{4.7}$$

$$\delta\rho_{\mathbf{k}}^a(t) = \int d\mathbf{r} e^{i\mathbf{k}\cdot\mathbf{r}} \delta\rho^a(\mathbf{r}, t) = \sum_i \exp\left[i\mathbf{k} \cdot \mathbf{r}_i^a(t) \right] - (2\pi)^3 \rho_0^a \delta(\mathbf{k}) \tag{4.8}$$

$\mathbf{J}_{\mathbf{k}}^a(t)$ is the Fourier component of the current of the ath atom in the solvent.

$$\mathbf{J}_{\mathbf{k}}^a(t) = \sum_i i\mathbf{k} \cdot \frac{\mathbf{p}_i^a}{m_a} \exp\left[i\mathbf{k} \cdot \mathbf{r}_i^a(t) \right] \equiv i\mathbf{k} \cdot \mathbf{j}_{\mathbf{k}}^a(t) \tag{4.9}$$

The correlation matrix $\mathbf{C} = (\mathbf{A}, \mathbf{A})$ is given by

$$(\mathbf{A}, \mathbf{A}) = \begin{pmatrix} (\Delta\mathbf{R}_\alpha, \Delta\mathbf{R}_\beta) & (\mathbf{P}_\alpha, \Delta\mathbf{R}_\beta) & (\delta\rho^a_{\mathbf{k}}, \Delta\mathbf{R}_\beta) & (\mathbf{J}^a_{\mathbf{k}}, \Delta\mathbf{R}_\beta) \\ (\Delta\mathbf{R}_\alpha, \mathbf{P}_\beta) & (\mathbf{P}_\alpha, \mathbf{P}_\beta) & (\delta\rho^a_{\mathbf{k}}, \mathbf{P}_\beta) & (\mathbf{J}^a_{\mathbf{k}}, \mathbf{P}_\beta) \\ (\Delta\mathbf{R}_\alpha, \delta\rho^b_{\mathbf{k}}) & (\mathbf{P}_\alpha, \delta\rho^b_{\mathbf{k}}) & (\delta\rho^a_{\mathbf{k}}, \delta\rho^b_{\mathbf{k}}) & (\mathbf{J}^a_{\mathbf{k}}, \delta\rho^b_{\mathbf{k}}) \\ (\Delta\mathbf{R}_\alpha, \mathbf{J}^b_{\mathbf{k}}) & (\mathbf{P}_\alpha, \mathbf{J}^b_{\mathbf{k}}) & (\delta\rho^a_{\mathbf{k}}, \mathbf{J}^b_{\mathbf{k}}) & (\mathbf{J}^a_{\mathbf{k}}, \mathbf{J}^b_{\mathbf{k}}) \end{pmatrix} \tag{4.10}$$

In Eq. (4.10), the following matrix elements vanish,

$$\begin{aligned} &(\Delta\mathbf{R}_\alpha, \mathbf{P}_\beta) = 0, \quad (\Delta\mathbf{R}_\alpha, \mathbf{J}^b_{\mathbf{k}}) = 0, \\ &(\mathbf{P}_\alpha, \Delta\mathbf{R}_\beta) = 0, \quad (\mathbf{P}_\alpha, \delta\rho^b_{\mathbf{k}}) = 0, \quad (\mathbf{P}_\alpha, \mathbf{J}^b_{\mathbf{k}}) = 0 \\ &(\delta\rho^a_{\mathbf{k}}, \mathbf{P}_\beta) = 0, \quad (\delta\rho^a_{\mathbf{k}}, \mathbf{J}^b_{\mathbf{k}}) = 0, \\ &(\mathbf{J}^a_{\mathbf{k}}, \Delta\mathbf{R}_\beta) = 0, \quad (\mathbf{J}^a_{\mathbf{k}}, \mathbf{P}_\beta) = 0, \quad (\mathbf{J}^a_{\mathbf{k}}, \delta\rho^b_{\mathbf{k}}) = 0 \end{aligned} \tag{4.11}$$

The matrix elements vanish because the following integrations becomes zero,

$$\int \mathrm{d}\mathbf{p}^{nN} \mathbf{p}^a_i \exp\left(-\beta \sum_i \sum_a \frac{\mathbf{p}^a_i \cdot \mathbf{p}^a_i}{2m_a} \right) = 0 \tag{4.12}$$

$$\int \mathrm{d}\mathbf{P}^{N_\mathrm{u}} \mathbf{P}_\alpha \exp\left(-\beta \sum_i \sum_a \frac{\mathbf{P}^2_\gamma}{2M_\gamma} \right) = 0 \tag{4.13}$$

Let us look at the non-vanishing elements. The integration concerning the correlation of momentum among solute atoms is performed readily to lead,

$$\begin{aligned} (\mathbf{P}_\alpha, \mathbf{P}_\beta) &= \frac{1}{Z_P} \int \mathrm{d}\mathbf{P}^{N_\mathrm{u}} \mathbf{P}_\alpha \mathbf{P}_\beta \exp\left(-\beta \sum_\gamma \mathbf{P}^2_\gamma / 2M_\gamma \right) \\ &= \frac{1}{Z_P} \int \mathrm{d}\mathbf{P}^{N_\mathrm{u}} \mathbf{P}_\alpha (-M_\beta k_\mathrm{B} T) \frac{\partial}{\partial \mathbf{P}_\beta} \exp\left(-\beta \sum_\gamma \mathbf{P}^2_\gamma / 2M_\gamma \right) = k_\mathrm{B} T M_\alpha \mathbf{1} \delta_{\alpha\beta} \end{aligned} \tag{4.14}$$

where $Z_P \equiv \int \mathrm{d}\mathbf{P}^{N_\mathrm{u}} \mathbf{P}_\alpha \mathbf{P}_\beta \exp\left[-\beta \sum_\gamma \mathbf{P}^2_\gamma / 2M_\gamma \right]$, and $\mathbf{1}$ is the unit (3×3) matrix. Equation (4.14) is the theorem known as the equipartition theorem in the statistical mechanics.

The general current–current correlation function $(\mathbf{J}^a_{\mathbf{k}}, \mathbf{J}^b_{\mathbf{k}})$ have non-vanishing correlation only between the same Cartesian components. Therefore, it is sufficient to define the current–current correlation function as

$$J_{ab}(\mathbf{k}) \equiv \frac{1}{N} \langle \mathbf{J}^a_{-\mathbf{k}} \cdot \mathbf{J}^b_{\mathbf{k}} \rangle \tag{4.15}$$

The detailed calculation is shown below,

$$
\begin{aligned}
J_{ab}(\mathbf{k}) &= \frac{1}{N}\sum_i\sum_j \frac{1}{m_a}\frac{1}{m_b}\left\langle \mathbf{p}_i^a \cdot \mathbf{p}_j^b \exp\left[-i\mathbf{k}\cdot\left(\mathbf{r}_i^a - \mathbf{r}_j^b\right)\right]\right\rangle \\
&= \frac{1}{N}\sum_i\sum_j \frac{1}{m_a}\frac{1}{m_b}\left\langle \mathbf{p}_i^a \cdot \mathbf{p}_j^b\right\rangle\left\langle \exp\left[-i\mathbf{k}\cdot\left(\mathbf{r}_i^a - \mathbf{r}_j^b\right)\right]\right\rangle \\
&= \frac{1}{N}\sum_i\sum_j \left\langle \mathbf{v}_i^a \cdot \mathbf{v}_j^b\right\rangle \delta_{ij}\left\langle \exp\left[-i\mathbf{k}\cdot\left(\mathbf{r}_i^a - \mathbf{r}_j^b\right)\right]\right\rangle \\
&= \frac{1}{N}\sum_i \left\langle \mathbf{v}_i^a \cdot \mathbf{v}_i^b\right\rangle\left\langle \exp\left[-i\mathbf{k}\cdot\left(\mathbf{r}_i^a - \mathbf{r}_i^b\right)\right]\right\rangle
\end{aligned} \tag{4.16}
$$

Rest of the elements, $(\Delta\mathbf{R}_\alpha, \Delta\mathbf{R}_\beta)$, $(\Delta\mathbf{R}_\alpha, \delta\rho_\mathbf{k}^b)$, $(\delta\rho_\mathbf{k}^a, \Delta\mathbf{R}_\beta)$, and $(\delta\rho_\mathbf{k}^a, \delta\rho_\mathbf{k}^b)$ are concerned only with the spatial coordinates. Let us consider them in order. The first element is the variance–covariance matrix of the displacement from equilibrium positions, or the structural fluctuation,

$$
L_{\alpha\beta} \equiv \left(\Delta\mathbf{R}_\alpha, \Delta\mathbf{R}_\beta\right) \tag{4.17}
$$

where $\mathbf{L}_{\alpha\beta}$ is a $(3N_\mathrm{u} \times 3N_\mathrm{u})$ matrix. The next element is $(\Delta\mathbf{R}_\alpha, \delta\rho_\mathbf{k}^b)$. Note that $(\Delta\mathbf{R}_\alpha, \delta\rho_\mathbf{k}^b) = \left\langle \Delta\mathbf{R}_\alpha \delta\rho_\mathbf{k}^b\right\rangle - (2\pi)^3\rho_0^b\delta(\mathbf{k}) = \left\langle \Delta\mathbf{R}_\alpha \delta\rho_\mathbf{k}^b\right\rangle$ since $\langle\Delta\mathbf{R}_\alpha\rangle = 0$. Therefore,

$$
\mathbf{B}_\mathbf{k}^{\alpha,b} \equiv \left(\Delta\mathbf{R}_\alpha, \delta\rho_\mathbf{k}^b\right) = \left\langle \Delta\mathbf{R}_\alpha, \delta\rho_\mathbf{k}^b\right\rangle \tag{4.18}
$$

It can be proved that the quantity and its transpose vanishes in the thermodynamic limit. (For the detailed proof, see Kim and Hirata (2013) in the reference.)

$$
\mathbf{B}_{\alpha,b}(\mathbf{k}) = \mathbf{B}_{a,\beta}(\mathbf{k}) = 0 \tag{4.19}
$$

The rest of the matrix element is the static structure factor of solvent, or the density pair correlation function, that is,

$$
\chi_{ab}(\mathbf{k}) \equiv \frac{1}{N}\left(\delta\rho_\mathbf{k}^a, \delta\rho_\mathbf{k}^b\right) = \frac{1}{N}\left(\delta\rho_{-\mathbf{k}}^a, \delta\rho_\mathbf{k}^b\right) \tag{4.20}
$$

Putting all the results together, we have the following matrix for $(\mathbf{A}, \mathbf{A})$,

$$
(\mathbf{A}, \mathbf{A}) = \begin{pmatrix}
L_{\alpha\beta} & \mathbf{O} & \mathbf{0} & \mathbf{0} \\
\mathbf{O} & k_\mathrm{B}TM_\alpha\mathbf{1}\delta_{\alpha\beta} & \mathbf{0} & \mathbf{0} \\
\mathbf{0}^\mathrm{T} & \mathbf{0}^\mathrm{T} & N\chi_{ab}(\mathbf{k}) & \underline{\mathbf{0}} \\
\mathbf{0}^\mathrm{T} & \mathbf{0}^\mathrm{T} & \underline{\mathbf{0}}^\mathrm{T} & N\mathbf{J}_{ab}(\mathbf{k})
\end{pmatrix} \tag{4.21}
$$

In Eq. (4.21), $\mathbf{O}$ represents the $(3N_\mathrm{u} \times 3N)$ zero matrix, $\mathbf{0}$ the $(3N_\mathrm{u} \times n)$ zero matrix, $\underline{\mathbf{0}}$ the $(n \times n)$ zero matrix, and the superscript T the transpose matrix.

Inverse of (A, A): The correlation matrix shown above is block-diagonal, therefore, it is easy to find the inverse $(\mathbf{A}, \mathbf{A})^{-1}$ as

$$(\mathbf{A}, \mathbf{A})^{-1} = \begin{pmatrix} \left(\mathbf{L}^{-1}\right)_{\alpha\beta} & \mathbf{O} & \mathbf{0} & \mathbf{0} \\ \mathbf{O} & (k_\mathrm{B}TM_\alpha)^{-1}\mathbf{1}\delta_{\alpha\beta} & \mathbf{0} & \mathbf{0} \\ \mathbf{0}^\mathrm{T} & \mathbf{0}^\mathrm{T} & N^{-1}\chi_{ab}(\mathbf{k}) & \underline{\mathbf{0}} \\ \mathbf{0}^\mathrm{T} & \mathbf{0}^\mathrm{T} & \underline{\mathbf{0}} & N^{-1}\mathbf{J}_{ab}^{-1}(\mathbf{k}) \end{pmatrix} \tag{4.22}$$

In the equation, the inverse matrices $(\mathbf{L}^{-1})_{\alpha\beta}$, $\chi_{\alpha\beta}^{-1}(k)$, and $J_{ab}^{-1}(k)$ are defined as

$$\sum_{\gamma=1}^{N_\mathrm{u}} L_{\alpha\gamma}\left(\mathbf{L}^{-1}\right)_{\gamma\beta} = \mathbf{1}\delta_{\alpha\beta} \quad \sum_{c=1}^{n} \chi_{ac}(\mathbf{k})\chi_{cb}^{-1}(\mathbf{k}) = \delta_{ab} \quad \sum_{c=1}^{n} J_{ac}(\mathbf{k})J_{cb}^{-1}(\mathbf{k}) = \delta_{ab} \tag{4.23}$$

Frequency Matrix: Now, we find the expression for the collective frequency matrix $i\Omega$ defined by

$$i\Omega_{\lambda\nu} \equiv \sum_{\lambda'} \left(\mathbf{A}_\lambda, \dot{\mathbf{A}}_{\lambda'}\right)\left[(\mathbf{A}, \mathbf{A})^{-1}\right]_{\lambda'\nu} \tag{4.24}$$

Let us look at the elements of the matrix $(\mathbf{A}, \dot{\mathbf{A}})$ first,

$$(\mathbf{A}, \dot{\mathbf{A}}) = \begin{pmatrix} \left(\Delta\mathbf{R}_\alpha, \Delta\dot{\mathbf{R}}_\beta\right) & \left(\mathbf{P}_\alpha, \Delta\dot{\mathbf{R}}_\beta\right) & \left(\delta\rho_\mathbf{k}^a, \Delta\dot{\mathbf{R}}_\beta\right) & \left(\mathbf{J}_\mathbf{k}^a, \Delta\dot{\mathbf{R}}_\beta\right) \\ \left(\Delta\mathbf{R}_\alpha, \dot{\mathbf{P}}_\beta\right) & \left(\mathbf{P}_\alpha, \dot{\mathbf{P}}_\beta\right) & \left(\delta\rho_\mathbf{k}^a, \dot{\mathbf{P}}_\beta\right) & \left(\mathbf{J}_\mathbf{k}^a, \dot{\mathbf{P}}_\beta\right) \\ \left(\Delta\mathbf{R}_\alpha, \delta\dot{\rho}_\mathbf{k}^b\right) & \left(\mathbf{P}_\alpha, \delta\dot{\rho}_\mathbf{k}^b\right) & \left(\delta\rho_\mathbf{k}^a, \delta\dot{\rho}_\mathbf{k}^b\right) & \left(\mathbf{J}_\mathbf{k}^a, \delta\dot{\rho}_\mathbf{k}^b\right) \\ \left(\Delta\mathbf{R}_\alpha, \dot{\mathbf{J}}_\mathbf{k}^b\right) & \left(\mathbf{P}_\alpha, \dot{\mathbf{J}}_\mathbf{k}^b\right) & \left(\delta\rho_\mathbf{k}^a, \dot{\mathbf{J}}_\mathbf{k}^b\right) & \left(\mathbf{J}_\mathbf{k}^a, \dot{\mathbf{J}}_\mathbf{k}^b\right) \end{pmatrix} \tag{4.25}$$

First, we obtain some elements of $\dot{\mathbf{A}}$ using the Liouville operator, or Eq. (4.2)

$$\Delta\dot{\mathbf{R}}_\alpha = i\hat{L}\Delta\mathbf{R}_\alpha = \mathbf{P}_\alpha / M_\alpha \tag{4.26}$$

$$\dot{\mathbf{P}}_\alpha = i\hat{L}\mathbf{P}_\alpha = \mathbf{F}_\alpha \tag{4.27}$$

$$\dot{\rho}_\mathbf{k}^a = i\hat{L}\rho_\mathbf{k}^a = i\mathbf{k} \cdot \mathbf{J}_\mathbf{k}^a \tag{4.28}$$

$$\dot{\mathbf{J}}_\mathbf{k}^a = \sum_i \frac{1}{m_a}\left(\dot{\mathbf{p}}_i^a + \mathbf{p}_i^a i\mathbf{k} \cdot \mathbf{J}_\mathbf{k}^a\right)\exp\left(i\mathbf{k} \cdot \mathbf{r}_i^a\right) \tag{4.29}$$

where $\mathbf{F}_\alpha$ is the total force exerted on the αth solute particle by the solvent as well as by other solute atoms. Actually, it is more convenient to use the integration by parts, when we compute the matrix elements concerning $\dot{\mathbf{P}}$ or $\dot{\mathbf{J}}^a_{\mathbf{k}}$. It is noted that whereas $\Delta\dot{\mathbf{R}}$ and $\dot{\rho}^a_{\mathbf{k}}$ involve single momentum ($\mathbf{P}$ or $\mathbf{p}_i$), $\dot{\mathbf{P}}$ and $\dot{\mathbf{J}}^a_{\mathbf{k}}$ include zero (since $\dot{\mathbf{p}}^a_i$ is the force acting on the ath atom of the ith molecule, which are only concerned with the position of solute atoms and solvent molecules), or two momentums (two $\mathbf{p}_i$). Using this fact, we can identify the vanishing elements,

$$\begin{array}{ccc} (\Delta\mathbf{R}_\alpha, \Delta\dot{\mathbf{R}}_\beta) = 0 & (\delta\rho^a_{\mathbf{k}}, \Delta\dot{\mathbf{R}}_\beta) = 0 & (\mathbf{J}^a_{\mathbf{k}}, \Delta\dot{\mathbf{R}}_\beta) = 0 \\ (\mathbf{P}_\alpha, \dot{\mathbf{P}}_\beta) = 0 & (\delta\rho^a_{\mathbf{k}}, \dot{\mathbf{P}}_\beta) = 0 & (\mathbf{J}^a_{\mathbf{k}}, \dot{\mathbf{P}}_\beta) = 0 \\ (\Delta\mathbf{R}_\alpha, \delta\dot{\rho}^b_{\mathbf{k}}) = 0 & (\mathbf{P}_\alpha, \delta\dot{\rho}^b_{\mathbf{k}}) = 0 & (\delta\rho^a_{\mathbf{k}}, \delta\dot{\rho}^b_{\mathbf{k}}) = 0 \\ (\Delta\mathbf{R}_\alpha, \dot{\mathbf{J}}^b_{\mathbf{k}}) = 0 & (\mathbf{P}_\alpha, \dot{\mathbf{J}}^b_{\mathbf{k}}) = 0 & (\mathbf{J}^a_{\mathbf{k}}, \dot{\mathbf{J}}^b_{\mathbf{k}}) = 0 \end{array} \tag{4.30}$$

The non-vanishing elements are

$$\begin{aligned} (\Delta\mathbf{R}_\alpha, \dot{\mathbf{P}}_\beta) &= -\frac{1}{M}(\mathbf{P}_\alpha, \mathbf{P}_\beta) = -k_{\mathrm{B}}T\mathbf{1}\delta_{\alpha\beta} \\ (\mathbf{P}_\alpha, \Delta\dot{\mathbf{R}}_\beta) &= \frac{1}{M}(\mathbf{P}_\alpha, \mathbf{P}_\beta) = k_{\mathrm{B}}T\mathbf{1}\delta_{\alpha\beta} \\ (\delta\rho^a_{\mathbf{k}}, \dot{\mathbf{J}}^b_{\mathbf{k}}) &= (\delta\dot{\rho}^a_{\mathbf{k}}, \mathbf{J}^b_{\mathbf{k}}) = i\mathbf{k}\cdot(\mathbf{J}^a_{\mathbf{k}}, \mathbf{J}^b_{\mathbf{k}}) = iN\mathbf{k}J_{ab}(k) \\ (\mathbf{J}^a_{\mathbf{k}}, \delta\dot{\rho}^b_{\mathbf{k}}) &= (\mathbf{J}^a_{\mathbf{k}}, \mathbf{J}^b_{\mathbf{k}})\cdot i\mathbf{k} = iN\mathbf{k}J_{ab}(k) \end{aligned} \tag{4.31}$$

Taking all these into account, we obtain

$$i\Omega = \begin{pmatrix} \mathbf{O} & k_{\mathrm{B}}T\mathbf{1}\delta_{\alpha\beta} & \mathbf{0} & \mathbf{0} \\ -k_{\mathrm{B}}T\mathbf{1}\delta_{\alpha\beta} & \mathbf{O} & \mathbf{0} & \mathbf{0} \\ \mathbf{0}^{\mathrm{T}} & \mathbf{0}^{\mathrm{T}} & \underline{\mathbf{0}} & iN\mathbf{k}J_{ab}(\mathbf{k}) \\ \mathbf{0}^{\mathrm{T}} & \mathbf{0}^{\mathrm{T}} & iN\mathbf{k}J_{ab}(\mathbf{k}) & \underline{\mathbf{0}} \end{pmatrix} \cdot (\mathbf{A}, \mathbf{A})^{-1} \tag{4.32}$$

Using the inverse correlation matrix (4.22), we get $i\Omega$ as

$$i\Omega = \begin{pmatrix} \mathbf{O} & \frac{1}{M_\alpha}\mathbf{1}\delta_{\alpha\beta} & \mathbf{0} & \mathbf{0} \\ -k_{\mathrm{B}}T\left(\mathbf{L}^{-1}\right)_{\alpha\beta} & \mathbf{O} & \mathbf{0} & \mathbf{0} \\ \mathbf{0}^{\mathrm{T}} & \mathbf{0}^{\mathrm{T}} & \underline{\mathbf{0}} & i\mathbf{k}\delta_{ab} \\ \mathbf{0}^{\mathrm{T}} & \mathbf{0}^{\mathrm{T}} & i\mathbf{k}\sum_{c=1}^{n} J_{ac}(\mathbf{k})\chi^{-1}_{cb}(\mathbf{k}) & \underline{\mathbf{0}} \end{pmatrix} \tag{4.33}$$

The Steady-State Part: As shown in Eq. (3.41) in Chap. 3, the *steady-state part* of the Langevin equation is given by $i\Omega\cdot\mathbf{A}(t)$. With Eq. (4.33), one finds

$$i\Omega \cdot \mathbf{A}(t) = \begin{pmatrix} \mathbf{P}_\alpha(t)/M_\alpha \\ -k_\mathrm{B}T\left(\mathbf{L}^{-1}\right)_{\alpha\beta} \cdot \Delta\mathbf{R}_\beta(t) \\ i\mathbf{k} \cdot \mathbf{J}^a_\mathbf{k}(t) \\ i\mathbf{k}\sum_{c=1}^{n} J_{ac}(\mathbf{k})\chi^{-1}_{cb}(\mathbf{k})\delta\rho^b_\mathbf{k}(t) \end{pmatrix} \tag{4.34}$$

The Fluctuating Force: The fluctuating force at $t = 0$, defined by Eq. (3.41), is given by

$$\mathbf{f} = \left(1 - \hat{P}\right)\dot{\mathbf{A}} = \dot{\mathbf{A}} - i\Omega\mathbf{A} \tag{4.35}$$

In the equation, we considered $\hat{P}\dot{\mathbf{A}} = (\mathbf{A}, \dot{\mathbf{A}}) \cdot (\mathbf{A}, \mathbf{A})^{-1}\mathbf{A} = i\Omega \cdot \mathbf{A}$. In order to find the expression for the fluctuational force, we first identify

$$\dot{\mathbf{A}} = \begin{pmatrix} \Delta\dot{\mathbf{R}}_\alpha \\ \dot{\mathbf{P}}_\alpha \\ \dot{\rho}^a_\mathbf{k} \\ \dot{\mathbf{J}}^a_\mathbf{k} \end{pmatrix} = \begin{pmatrix} \mathbf{P}_\alpha/M_\alpha \\ \mathbf{F}_\alpha \\ i\mathbf{k} \cdot \mathbf{j} \\ \dot{\mathbf{J}}^a_\mathbf{k} \end{pmatrix} \tag{4.36}$$

and

$$i\Omega \cdot \mathbf{A} = \begin{pmatrix} \mathbf{P}_\alpha/M_\alpha \\ -k_\mathrm{B}T\left(\mathbf{L}^{-1}\right)_{\alpha\beta} \cdot \Delta\mathbf{R}_\beta \\ i\mathbf{k} \cdot \mathbf{J}^a_\mathbf{k} \\ i\mathbf{k}\sum_{c=1}^{n} J_{ac}(\mathbf{k})\chi^{-1}_{cb}(\mathbf{k})\delta\rho^b_\mathbf{k} \end{pmatrix} \tag{4.37}$$

which are obtained by setting $t = 0$ in Eq. (4.34). Putting Eqs. (4.35)–(4.37) all together, one finds the expression for the fluctuating force as

$$\mathbf{f} = \begin{pmatrix} 0 \\ \mathbf{W}_\alpha \\ 0 \\ \Xi^a_\mathbf{k} \end{pmatrix} \quad \text{and} \quad \mathbf{f}(t) = \exp\left[it(1-\hat{P})\hat{L}\right]\begin{pmatrix} 0 \\ \mathbf{W}_\alpha \\ 0 \\ \Xi^a_\mathbf{k} \end{pmatrix} \tag{4.38}$$

where

$$\begin{aligned} \mathbf{W}_\alpha &= \mathbf{F}_\alpha + k_\mathrm{B}T\sum_\beta \left(\mathbf{L}^{-1}\right)_{\alpha\beta} \cdot \Delta\mathbf{R}_\beta \\ \Xi^a_\mathbf{k} &= \dot{\mathbf{J}}^a_\mathbf{k} - i\mathbf{k}\sum_{b,c} J_{ac}(k)\chi^{-1}_{cb}(k)\delta\rho^b_\mathbf{k} \end{aligned} \tag{4.39}$$

The Memory Matrix: The memory function matrix $\zeta(t)$ is calculated as

$$\mathbf{K}(t) \equiv \frac{(\mathbf{f}, \mathbf{f}(t))}{(\mathbf{A}, \mathbf{A})}$$
$$= \begin{pmatrix} \mathbf{O} & \mathbf{O} & \mathbf{0} & \mathbf{0} \\ \mathbf{O} & \frac{1}{k_\mathrm{B} T M_\alpha}(\mathbf{W}_\alpha, \mathbf{W}_\beta(t)) & \mathbf{0} & \frac{1}{N}\sum_b J_{ab}^{-1}(k)\left(\Xi_\mathbf{k}^b, W_\beta(t)\right) \\ \mathbf{0}^\mathrm{T} & \mathbf{0}^\mathrm{T} & \mathbf{0} & \mathbf{0} \\ \mathbf{0}^\mathrm{T} & \frac{1}{k_\mathrm{B} T M_\alpha}(\mathbf{W}_\alpha, \Xi_\mathbf{k}^b(t)) & \mathbf{0} & \frac{1}{N}\sum_c J_{ac}^{-1}(k)\left(\Xi_\mathbf{k}^c, \Xi_\mathbf{k}^b(t)\right) \end{pmatrix} \tag{4.40}$$

where $\mathbf{W}(t)$ and $\Xi_\mathbf{k}^a$ are defined as

$$\mathbf{W}(t) \equiv \mathrm{e}^{it(1-\hat{P})\hat{L}}\mathbf{W} \quad \text{and} \quad \Xi_\mathbf{k}^a(t) \equiv \mathrm{e}^{it(1-\hat{P})\hat{L}}\Xi_\mathbf{k}^a \tag{4.41}$$

The two terms in Eq. (4.40) are explicitly dependent on N. In the thermodynamic limit in which N is taken to be infinite while the number of atoms N_u in a solute remains finite, the term $N^{-1}\sum_b J_{ab}^{-1}(k)(\Xi_\mathbf{k}^b, \mathbf{W}(t))$ vanishes since the ensemble average $(\Xi_\mathbf{k}^b, \mathbf{W}(t))$ remains finite. The other term $N^{-1}\sum_c J_{ac}^{-1}(k)(\Xi_\mathbf{k}^c, \Xi_\mathbf{k}^b(t))$ will not vanish since the ensemble average $(\Xi_\mathbf{k}^c, \Xi_\mathbf{k}^b(t))$ is proportional to N. Therefore, only the latter term survives in the thermodynamic limit. It can be also proved that

$$(\mathbf{W}_\alpha, \Xi_k^b(t)) = 0 \tag{4.42}$$

Therefore, the final expression of the memory matrix is given by

$$\mathbf{K}(t) = \begin{pmatrix} \mathbf{O} & \mathbf{O} & \mathbf{0} & \mathbf{0} \\ \mathbf{O} & \frac{1}{k_\mathrm{B} T M_\alpha}(\mathbf{W}_\alpha, \mathbf{W}_\beta(t)) & \mathbf{0} & \mathbf{0} \\ \mathbf{0}^\mathrm{T} & \mathbf{0}^\mathrm{T} & \mathbf{0} & \mathbf{0} \\ \mathbf{0}^\mathrm{T} & \mathbf{0}^\mathrm{T} & \mathbf{0} & \frac{1}{N}\sum_c J_{ac}^{-1}(k)\left(\Xi_\mathbf{k}^c, \Xi_\mathbf{k}^b(t)\right) \end{pmatrix} \tag{4.43}$$

Generalized Langevin Equations for a System of Protein in Solution: Putting all the results derived from Eq. (4.34) through Eq. (4.43) together into the generalized Langevin Eq. (3.41), we are now ready to write down explicit formulas for the structural dynamics of a protein conjugated with the density fluctuation of solvent.

$$\frac{\mathrm{d}\Delta\mathbf{R}_\alpha(t)}{\mathrm{d}t} = \frac{\mathbf{P}_\alpha(t)}{M_\alpha} \tag{4.44}$$

$$\frac{\mathrm{d}\mathbf{P}_\alpha(t)}{\mathrm{d}t} = -k_\mathrm{B}T\sum_\beta \left(\mathbf{L}^{-1}\right)_{\alpha\beta} \cdot \Delta\mathbf{R}_\beta(t) - \int_0^t \mathrm{d}s \sum_\beta \Gamma_{\alpha\beta}(t-s) \cdot \frac{\mathbf{P}_\beta(s)}{M_\beta} + \mathbf{W}_\alpha(t) \tag{4.45}$$

$$\frac{\mathrm{d}\delta\rho_{\mathbf{k}}^{a}(t)}{\mathrm{d}t} = i\mathbf{k}\cdot\mathbf{J}_{\mathbf{k}}^{a}(t) \tag{4.46}$$

$$\frac{\mathrm{d}\mathbf{J}_{\mathbf{k}}^{a}(t)}{\mathrm{d}t} = i\mathbf{k}\sum_{c=1}^{n} J_{ac}(k)\chi_{cb}^{-1}(k)\delta\rho_{\mathbf{k}}^{b} - \frac{1}{N}\sum J_{ac}^{-1}(k)\int_{0}^{t}\mathrm{d}s\mathbf{M}_{\mathbf{k}}^{cb}(t-s)\cdot\mathbf{J}_{\mathbf{k}}^{b}(s) + \Xi_{\mathbf{k}}^{a}(t) \tag{4.47}$$

In the above set of dynamic equations for the solute and solvent molecules, the random forces take the following forms:

$$\mathbf{W}_{\alpha} = \exp\left[it(1-\hat{P})\hat{L}\right]\left(\mathbf{F}_{\alpha} + k_{\mathrm{B}}T\sum_{\beta}\left(\mathbf{L}^{-1}\right)_{\alpha\beta}\cdot\Delta\mathbf{R}_{\beta}\right) \tag{4.48}$$

$$\Xi_{\mathbf{k}}^{a} = \exp\left[it(1-\hat{P})\hat{L}\right]\left(\dot{\mathbf{J}}_{\mathbf{k}}^{a} - i\mathbf{k}\sum_{b,c} J_{ac}(k)\chi_{cb}^{-1}(k)\delta\rho_{\mathbf{k}}^{b}\right) \tag{4.49}$$

The memory functions in Eqs. (4.45) and (4.47) are given by the time correlations of the random forces,

$$\Gamma_{\alpha\beta}(t) = \frac{1}{k_{\mathrm{B}}T}\left\langle\mathbf{W}_{\alpha}(t)\mathbf{W}_{\beta}(0)\right\rangle \tag{4.50}$$

$$\mathbf{M}_{\mathbf{k}}^{bc}(t) = \left\langle\Xi_{\mathbf{k}}^{b}(t)\Xi_{-\mathbf{k}}^{c}(0)\right\rangle \tag{4.51}$$

Physics of the Generalized Langevin Equations: The first terms in the right-hand-side in Eqs. (4.45) and (4.47) represent the force to restore the atoms from a fluctuated or a non-equilibrium state to the equilibrium one. It should be noted that the restoring force is *proportional* or *linear* to the fluctuation or the displacement itself in the both equations. It is important to note that no such assumptions for the harmonicity or linearity of the force are made in the course of derivation of the equations. The harmonicity or linearity is nothing but a consequence of the projection of all the degrees of freedom with respect to water molecules on to the four dynamic variables defined by Eq. (4.5). Thus, it was proved that both the structural fluctuation of a biomolecule in water and the density fluctuation of water around a biomolecule are *harmonic* in its nature. The finding is extremely important, because it lets us use all the mathematical tools developed for the harmonic analysis, such as the *linear algebra*.

4.2 GLE for a Biomolecule in Water

The generalized Langevin equations for a biomolecule in water look very similar to the phenomenological Langevin equation for a harmonic oscillator in viscous fluid, analyzed originally by Wang and Uhlenbeck, and later by Lamb and Szabo in their Langevin mode analysis (Wang and Uhlenbeck 1945; Lamm and Szabo 1986).

Let us put Eqs. (4.44) and (4.45) together to lead more familiar form of the Langevin equation,

$$M_\alpha \frac{d^2 \Delta \mathbf{R}_\alpha(t)}{dt^2} = -k_B T \sum_\beta \left(\mathbf{L}^{-1}\right)_{\alpha\beta} \cdot \Delta \mathbf{R}_\beta(t) - \int_0^t ds \sum_\beta \Gamma_{\alpha\beta}(t-s) \cdot \frac{\mathbf{P}_\beta(s)}{M_\beta} + \mathbf{W}_\alpha(t). \tag{4.52}$$

The three terms in the right-hand side represent forces acting on atoms in the biomolecule. The last term is the random or fluctuational force, while the second term represents the frictional force acting on atoms in protein. Those forces have their origin essentially in the thermal motion of solvent molecules, and they have been treated in the phenomenological Langevin theory in one way or the other.

It is the first term that was not identified in the phenomenological theories. The first term in the right-hand-side in the equation represents the force to restore the atoms from a fluctuated or a non-equilibrium state to the equilibrium state. It should be noted that the restoring force is proportional to the fluctuation or the displacement itself. So, the forces are formally identical to those acting in the *harmonic oscillator*. It is important to note that no such assumptions for the harmonicity or linearity of the force are made in the course of derivation of the equation. The harmonicity or linearity is nothing but a consequence of the *projection* of all the degrees of freedom with respect to water molecules onto the two dynamic variables concerning the biomolecule. Thus, the equation *proves* that the structural fluctuation of protein in water is *harmonic* in its nature. The finding is extremely important from a theoretical view point, because it allows us to use all the mathematical tools developed for the harmonic analysis, such as the *linear algebra*, in order to solve complicated problem related to the structural fluctuation of a biomolecule in water.

The structural fluctuation of a biomolecule in water is apparently extremely complicated, involving a wide range of oscillatory modes from a vibrational motion of a local bond to the collective motion extending over the entire molecule. The harmonicity or linearity of the structural fluctuation allows us to decouple the composite oscillation into independent oscillatory modes, which have a hierarchical order in terms of the frequency.

The linearity or the harmonicity of the structural fluctuation is advantageous from a view point of the computational chemistry, because the computational programs for the linear algebra including matrix diagonalization are well developed in the field of quantum chemistry. One may take advantage of the technologies developed in the field of quantum chemistry, such as the *Gaussian* program to evaluate the electronic structure of a molecule.

In order to make the physics of the equation more transparent, let us rewrite the equation without the last two terms representing the frictional force and the random force.

$$M_\alpha \frac{d^2 \Delta \mathbf{R}_\alpha(t)}{dt^2} = -\sum_\beta K_{\alpha\beta} \Delta \mathbf{R}_\beta(t) \tag{4.53}$$

$$K_{\alpha\beta} \equiv k_B T \left(\mathbf{L}^{-1}\right)_{\alpha\beta} \tag{4.54}$$

where $\left(\mathbf{L}^{-1}\right)_{\alpha\beta}$ represents the (α, β) element of the inverse of the variance–covariance matrix, $\mathbf{L}$, defined by Eq. (4.17), or $L_{\alpha\beta} \equiv \langle \Delta\mathbf{R}\Delta\mathbf{R}\rangle_{\alpha\beta}$. The equation is formally identical to the system of a harmonic oscillator in which all the atoms are connected by springs, the *force constant* or *Hessian* of which is $K_{\alpha\beta}$. So, the equation describes a stationary or a steady-state fluctuation in which atoms are in a harmonic motion around an equilibrium position. Then, how the equilibrium position and the force constant of the harmonic motion is defined?

In the case of an harmonic oscillator placed in *vacuum*, the equilibrium is defined by the structure that minimizes the mechanical potential-energy $U(\{\mathbf{R}\})$ of the molecule,

$$\frac{\partial U(\{\mathbf{R}\})}{\partial \Delta \mathbf{R}_\alpha} = 0 \quad \text{(a biomolecule in vacuum)} \tag{4.55}$$

The force constant is defined by the second derivative of $U(\{\mathbf{R}\})$ with respect to the atomic coordinates of the molecule as,

$$K_{\alpha\beta} \equiv \frac{\partial^2 U(\{\mathbf{R}\})}{\partial \Delta \mathbf{R}_\alpha \partial \Delta \mathbf{R}_\beta} \quad \text{(a biomolecule in vacuum)} \tag{4.56}$$

It will be worthwhile to note that such harmonic analysis for a biomolecule in vacuum is valid only in the case that the Taylor expansion of U can be truncated in the second order of the displacement, $\Delta\mathbf{R}$.

In the case of a real biomolecule in water, the thermodynamic stability of a structure is determined by the *free energy*, or the *potential of mean force*. Therefore, it will be natural to define the equilibrium structure by the one that makes the *free energy* minimum, and the *force constant* by the second derivative of the free energy with respect to the atomic coordinates of the biomolecule.

$$\frac{\partial F(\{\mathbf{R}\})}{\partial \Delta \mathbf{R}_\alpha} = 0 \quad \text{(a biomolecule in water)} \tag{4.57}$$

$$K_{\alpha\beta} \equiv \frac{\partial^2 F(\{\mathbf{R}\})}{\partial \Delta \mathbf{R}_\alpha \partial \Delta \mathbf{R}_\beta} \quad \text{(a biomolecule in water)} \tag{4.58}$$

where the structural free energy or the potential of mean force is defined by

$$F(\{\mathbf{R}\}) \equiv U(\{\mathbf{R}\}) + \Delta\mu(\{\mathbf{R}\}) \tag{4.59}$$

In the definition, $U(\{\mathbf{R}\})$ denotes the mechanical potential-energy among atoms in the biomolecule, while $\Delta\mu(\{\mathbf{R}\})$ is the chemical potential of the biomolecule in water, which can be evaluated by means of the statistical mechanics of liquids, such as the RISM/3D-RISM theory.

The physics attributed above leads to an important conclusion concerning the probability distribution of the structure of protein in water. By integrating Eq. (4.58) with respect to atomic coordinates of protein, $\Delta\mathbf{R}_\alpha$ and $\Delta\mathbf{R}_\beta$, one finds the free energy of protein in an equilibrium conformation as,

$$F(\{\Delta\mathbf{R}\}) = \frac{1}{2} k_B T \sum_{\alpha,\beta} \Delta\mathbf{R}_\alpha K_{\alpha\beta} \Delta\mathbf{R}_\beta \tag{4.60}$$

Then, the probability distribution of the structural fluctuation $(\Delta\mathbf{R}_1, \Delta\mathbf{R}_2, \ldots, \Delta\mathbf{R}_N)$ may be expressed as,

$$\begin{aligned} w(\{\Delta\mathbf{R}\}) &= C \exp[-F(\{\Delta\mathbf{R}\})/k_B T] \\ &= \sqrt{\frac{\|K\|}{(2\pi)^{3N}}} \exp\left[-\frac{1}{2}\sum_\alpha \sum_\beta \Delta\mathbf{R}_\alpha K_{\alpha\beta} \Delta\mathbf{R}_\beta\right], \end{aligned} \tag{4.61}$$

where $\|K\|$ denotes the determinant of the matrix, an element of which is $K_{\alpha\beta}$ defined by Eq. (4.58). The equation implies that the probability distribution of the structural fluctuation is *Gaussian*, characterized by the half-width, or the variance–covariance matrix, defined by Eq. (4.17) or $L_{\alpha\beta} \equiv \langle \Delta\mathbf{R}\Delta\mathbf{R}\rangle_{\alpha\beta}$.

Experimental Evidence for the Gaussian Fluctuation: An experimental evidence of the Gaussian fluctuation of a biomolecule in water is provided by the so-called "Guinier plot" concerning the small angle X-ray scattering (SAXS) as been already mentioned in Chap. 1 (Introduction) (Kataoka et al. 1995). The plot includes the information of the structural fluctuation in terms of the mean square displacement of atoms in the molecule.

The mean square displacement of a biomolecule is defined by,

$$M = \sum_\alpha^{N_u} \langle \Delta\mathbf{R}_\alpha^2 \rangle, \tag{4.62}$$

where α labels all atoms in protein, and N_u denotes the number of atoms. According to the Kim–Hirata theory described above, the probability distribution of the displacement of atoms is defined by,

$$w(\{\Delta\mathbf{R}\}) = (2\pi L_{\alpha\alpha})^{-3N_u/2} \exp\left[-\frac{1}{2}\sum_{\alpha}^{N_u}\frac{\Delta\mathbf{R}_\alpha^2}{L_{\alpha\alpha}}\right], \tag{4.63}$$

where $L_{\alpha\alpha} \equiv \langle \Delta\mathbf{R}_\alpha \Delta\mathbf{R}_\alpha \rangle$. The structure factor ($S(Q, \omega = 0)$) is defined by the Fourier transform of $w(\{\Delta\mathbf{R}\})$ into the wavenumber space. In order to perform the transformation, we define a displacement vector in the three dimensional coordinate space by

$$\Delta\mathbf{R} = \sum_{\alpha} \Delta\mathbf{R}_\alpha \tag{4.64}$$

and its differential volume element by

$$d\Delta\mathbf{R} = \prod_{\alpha} d\Delta\mathbf{R}_\alpha \tag{4.65}$$

The structure factor, then, may be defined by the Fourier transform of $w(\{\Delta\mathbf{R}_\alpha\})$,

$$S(Q, \omega = 0) \equiv \int w(\Delta\mathbf{R}) \exp(-i\mathbf{Q}\cdot\Delta\mathbf{R}) \mathrm{d}\Delta\mathbf{R}. \tag{4.66}$$

With the definitions Eqs. (4.63) and (4.65) in mind, Eq. (4.66) may be written as,

$$S(Q, \omega = 0) = \prod_{\alpha}^{N_u} (2\pi L_{\alpha\alpha})^{-3/2} \int \exp\left[-\frac{\Delta R_\alpha^2}{2L_{\alpha\alpha}}\right] \exp(-i\mathbf{Q}\cdot\Delta\mathbf{R}) \mathrm{d}\Delta\mathbf{R}_\alpha \tag{4.67}$$

The Fourier transform of the Gaussian function can be readily performed to give,

$$S(Q, \omega = 0) = \prod_{\alpha}^{N_u} \exp\left(-\frac{L_{\alpha\alpha}}{2}Q^2\right), \tag{4.68}$$

or

$$\log S(Q, \omega = 0) = -\frac{1}{2}\left(\sum_{\alpha}^{N_u} L_{\alpha\alpha}\right)Q^2 = -\frac{1}{2}MQ^2 \tag{4.69}$$

The derivation proves that $\log S(Q, \omega)$ is proportional to Q^2, and the proportional constant is the mean square displacement M defined by Eq. (4.62).

As been shown in Fig. 1.4 in Chap. 1, the Guinier plot exhibits clearly the behavior predicted by Eq. (4.69).

4.3 Temporal Density Fluctuation of Water Around a Biomolecule

Equations (4.46) and (4.47) describe the temporal density fluctuation of water around protein. The equations are readily merged into the following equation.

$$\begin{aligned}\frac{\partial^2 \delta\rho_a(\mathbf{r},t)}{\partial t^2} &= -\sum_{b,c}\int \kappa_{ab}^{\mathrm{solv}}(\mathbf{r},\mathbf{r}')\delta\rho_c(\mathbf{r}',t)\mathrm{d}\mathbf{r}' \\ &+ \left(\frac{1}{N}\sum_{b,c}\int \mathrm{d}\mathbf{r}'\int \mathrm{d}\mathbf{r}'' J_{ab}(\mathbf{r},\mathbf{r}')\int_0^t \mathrm{d}sM_{bc}\right. \\ &\left.(\mathbf{r}';t-s)\frac{\mathrm{d}\delta\rho_c(\mathbf{r}'',s)}{\mathrm{d}t}\mathrm{d}s\right) - \Xi_a(\mathbf{r},t), \end{aligned}\tag{4.70}$$

where $\kappa_{ab}^{\mathrm{solv}}(\mathbf{r},\mathbf{r}')$ is defined by,

$$\kappa_{ab}^{\mathrm{solv}}(\mathbf{r},\mathbf{r}') \equiv -\sum_c \int \mathrm{d}\mathbf{r}'' J_{ac}(\mathbf{r},\mathbf{r}'')\chi_{cb}^{(2)}(\mathbf{r}'',\mathbf{r}')^{-1} \tag{4.71}$$

J_{ac} in the equation represents the kinetic contribution of molecules to the density fluctuation, which is defined by,

$$J_{ac}(\mathbf{r},\mathbf{r}') = \sum_i \left\langle \mathbf{v}_i^a \cdot \mathbf{v}_i^c \right\rangle \omega_{ac}\left(\left|\mathbf{r}-\mathbf{r}'\right|\right). \tag{4.72}$$

In the equation, $\mathbf{v}_i^a$ and $\mathbf{v}_i^b$ denote the velocity of atoms a and b belonging to the same molecule i, and ω_{ab} represents the *intra*molecular correlation function defined by

$$\begin{aligned}\omega_{ab}\left(\left|\mathbf{r}-\mathbf{r}'\right|\right) &= \rho\delta_{ab}\delta\left(\left|\mathbf{r}-\mathbf{r}'\right|\right) + \rho(1-\delta_{ab})y_{ab}\left(\left|\mathbf{r}-\mathbf{r}'\right|\right) \\ y_{ab}\left(\left|\mathbf{r}-\mathbf{r}'\right|\right) &= \frac{1}{4\pi l_{ab}^2}\delta\left(\left|\mathbf{r}-\mathbf{r}'\right| - l_{ab}\right)\end{aligned}\tag{4.73}$$

In the equations, l_{ab} is the distance, or "bond", between atoms a and b in a same molecule, δ_{ab} is the Kronecker delta-function, and $\delta\left(\left|\mathbf{r}-\mathbf{r}'\right|\right)$ is the Dirac delta-function, respectively. (See Eq. (2.47) in Chap. 2 for details. Notations are slightly changed for the consistency of the description in the chapter.) The molecular structure of solvent is defined by providing l_{ab} for all pairs of atoms belonging to the molecule. $\chi_{ab}^{(2)}$ is the standard deviation, or dispersion, of the density fluctuation around a biomolecule.

$$\chi_{ab}^{(2)}(\mathbf{r}, \mathbf{r}') \equiv \left\langle \delta\rho_a(\mathbf{r})\delta\rho_b(\mathbf{r}')\right\rangle \tag{4.74}$$

In the equation, $\delta\rho_a(\mathbf{r})$ and $\delta\rho_b(\mathbf{r}')$ are the density fluctuation of atom a and b of water molecules around the biomolecule. So, $\chi_{ab}^{(2)}$ is nothing but the density pair correlation function. However, it should be noted that it is not the same quantity as that in pure water, because each atom in the pair is subject to the inhomogeneous field exerted by protein atoms.

The time average of the density fluctuation, $\langle \delta\rho_a(\mathbf{r})\rangle$, is the equilibrium density correlation function of atom a of solvent around the protein, which can be calculated by means of the RISM/3D-RISM theory.

The equation for the *steady-state* fluctuation of the density around a biomolecule may be obtained by ignoring the second and third terms of Eq. (4.70) as,

$$\frac{\partial^2 \delta\rho_a(\mathbf{r}, t)}{\partial t^2} = -\sum_b \int \kappa_{ab}^{\text{solv}}(\mathbf{r}, \mathbf{r}')\delta\rho_b(\mathbf{r}', t)d\mathbf{r}' \tag{4.75}$$

The accelerating force driving the density fluctuation is again *linear* to the density fluctuation.

The mathematical isomorphism with a harmonic oscillator is completed by defining the free energy of solvent as,

$$F_{\text{solv}} \equiv \int\int \mu_{\text{solv}}(\mathbf{r}, \mathbf{r}')d\mathbf{r}d\mathbf{r}' \tag{4.76}$$

In the equation, the *local free energy* $\mu_{\text{solv}}(\mathbf{r}, \mathbf{r}')$ at a pair of positions $\mathbf{r}$ and $\mathbf{r}'$ is defined by

$$\mu_{\text{solv}}(\mathbf{r}, \mathbf{r}') \equiv \frac{1}{2}k_B T \sum_a \sum_b \delta\rho_a(\mathbf{r})\kappa_{ab}^{\text{solv}}(\mathbf{r}, \mathbf{r}')\delta\rho_b(\mathbf{r}') \tag{4.77}$$

In the definition, $\kappa_{ab}^{\text{solv}}(\mathbf{r}, \mathbf{r}')$ is identified as the second (functional) derivative of the free energy F_{solv} with respect to the density fluctuation at fixed positions $\mathbf{r}$ and $\mathbf{r}'$,

$$\kappa_{ab}^{\text{solv}}(\mathbf{r}, \mathbf{r}') = \frac{\delta^2(F_{\text{solv}}/k_B T)}{\delta\rho_a(\mathbf{r})\delta\rho_b(\mathbf{r}')}. \tag{4.78}$$

The identification is similar to the density-functional theory of liquids, except for the kinetic factor defined by Eq. (4.72) (Hansen and Mcdonald 1986). The reason why the kinetic factor is included in Eq. (4.71) is because the momentum density or flux also contributes to change of the free energy.

The physics of Eq. (4.75) is interpreted as follows in analogy to a *harmonic oscillator*. The *force* to restore the *equilibrium density* of solvent-atom, a, at position $\mathbf{r}$ is linear to the density fluctuation of solvent-atom, b, at position $\mathbf{r}'$. The factor $\kappa_{ab}^{\text{solv}}(\mathbf{r}, \mathbf{r}')$ in Eq. (4.75) is the *force constant* of the *restoring force*, that is the second derivative of the free energy with respect to the density fluctuation of atoms a and b, as is expressed by Eq. (4.78).

The harmonic analysis of the density fluctuation in pure water, not in the vicinity of a biomolecule, has been shown in the preceding chapter, based on the Fourier space counterpart of Eq. (4.71), or,

$$\langle \omega_k^2 \rangle_{ab} = k^2 \sum_c J_{ac}(k) \left(\chi(k)^{-1} \right)_{cb}$$

The harmonic analysis involves the diagonalization of the matrix, $\langle \omega_k^2 \rangle_{ab}$, which is a (3×3) matrix, corresponding to the O, H, H atoms in a water molecule.

4.3.1 Entropy of Fluctuation in Biomolecular Solutions

Entropy is a thermodynamic variable which reflects a variety of *fluctuations* taking place in biomolecular systems most directly. In the present section, we derive mathematical formula to evaluate entropy that is related to the two types of fluctuations discussed in the preceding sections: the conformational or structural fluctuation of a biomolecule in water, and the density fluctuation of water around a biomolecule. Before doing so, we briefly review the theoretical studies of the entropy carried out during the past few decades.

Entropy in solutions is categorized into two groups, depending on whether molecules in the system are interacting with each other (Prigogine and Defay 1966; Rowlinson 1969). The entropy which is not related to the intermolecular interactions includes the *ideal* and *mixing* entropies. The ideal entropy has its origin in the kinetics of individual molecules in the system. The entropy is a single molecular property, which can be calculated readily from the classical mechanics of a molecule, as a function of the mass, the moment of inertia, and the force constant (McQuarrie 1976).

The mixing entropy is a quantity to measure how well the species in solution is mixed. It depends just on the concentration of the system, defined as,

$$S_{\text{m}} = -\sum_i k_{\text{B}} \log x_i,$$

where x_i denotes the concentration (*mol fraction*) of species i in the solution (Prigogine and Defay 1966).

Our contemporary interest concerning entropy is the one related to the interaction among atoms and molecules. Such entropy is referred to as *excess entropy*. There are two types of the excess entropy that is intimately related to fluctuations in biomolecular systems: conformational entropy of a biomolecule in water and the entropy of water around a biomolecule.

Conformational Entropy of a Biomolecule: The entropy is a thermodynamic quantity that reflects faithfully the magnitude of temporal as well as structural fluctuations (Go et al. 1968, 1970; Karplus and Kushick 1981; Levy et al. 1981; Lazaridis and Karplus 1996; Chong and Ham 2011, 2013, 2016). As been well documented, entropy is an essential factor to determine the conformational stability of protein in water. The reason is because the stability of *denatured state* of protein is dominated largely by the conformational entropy of protein (Levinthal 1969; Anfinsen 1973; Bryngelson et al. 1995).

The formulation of the structural entropy of a biomolecule has been made originally by Go, Go, and Scheraga based on the statistical-mechanical (Go et al. 1968, 1970). In the formulation, the authors have calculated the conformational entropy of protein based on the Gibbs formula of entropy, assuming that the probability distribution of the dihedral angles is *Gaussian*. The attempt was not very successful, unfortunately, because the model they have used was not so realistic. The most serious defect of the treatment is the model they have used for protein, which is a protein *in vacuum*. Two problems should be pointed out for using the model, *protein in vacuum*, to calculate the structural entropy of a biomolecule. At first, the equilibrium structure of a biomolecule placed *in vacuum* is completely different from that in water. Secondly, the variance–covariance matrix of the structural fluctuation is entirely different from the real protein in water. It is worthwhile to note that the *Gaussian* distribution of the structural fluctuation is resulted from the truncation of the interaction potential-energy at the second order of the Taylor expansion with respect to the fluctuation of atoms from equilibrium.

The calculation of the structural entropy of protein has been put on the foundation of the molecular-dynamics simulation by Karplus and his coworkers (Karplus et al. 1981; Levy et al. 1981). They have analyzed the variance–covariance matrix of a protein over the MD trajectory to calculate the conformational entropy. The method has been applied to more realistic systems in which a solute molecule is immersed in water (Schlitter 1993). Unfortunately, the entropy calculation has never converged to the experimental result. Two reasons are conceivable for the disagreement. At first, the dynamics may be evolving on a *potential* energy surface, all for solute, solvent, and solute–solvent interactions, which is *non-quadratic*. Therefore, the probability distribution of the structural fluctuation is not strictly the *Gaussian* distribution. If one cools down the system to the potential-energy minimum, just as the case in the normal mode analysis, it may satisfy the Gaussian distribution. However, such an attempt may ruin the original purpose to calculate the conformational entropy itself, since the system cooled down to zero degree of temperature is not just a *material* without life. The other reason for the disagreement is that the method can never reach the two limits of statistical mechanics, that is, thermodynamic as well as Ergodic limits. If the method meets the two limits, the conformational fluctuation should become strictly *Gaussian* due to the *central limiting theorem*.

Recently, a new computational method to estimate the conformational entropy of protein in water have been proposed by Chong and Ham based on the MD simulation combined with the RISM/3D-RISM theory (Chong and Ham 2011, 2013). In the calculation, the driving force of protein atoms is calculated as the first derivative of the free energy surface, or the potential of mean force (PMF), of protein, which incorporates the contribution from the solvation free energy. The PMF thus calculated should be *quadratic* with respect to the atomic coordinates of protein at least in good approximation, as was shown in Eq. (4.60). Then, the authors have performed the cumulant expansion of the PMF of solute, the structure of which is fluctuating with certain probability. The second order of expansion is expected to give the conformational entropy of a protein as

$$S_{\mathrm{conf}} = \frac{k_{\mathrm{B}}}{2}\left\langle (F - \langle F \rangle)^2 \right\rangle \tag{4.79}$$

F in Eq. (4.79) is PMF or the free energy of solute, defined by $F \equiv E + \Delta\mu$. In Eq. (4.79), E denotes the interatomic potential-energy of protein, $\Delta\mu$ represents the solvation free energy, and $\langle F \rangle$ is the average of F over the MD trajectory. The authors have calculated E and $\Delta\mu$ along the trajectory of the MD simulation of protein in aqueous solution. In order to calculate $\Delta\mu$, they have employed the RISM/3D-RISM theory. An important observation made by the authors is that the probability distribution of the free-energy fluctuation in fact shows a *Gaussian distribution*. The observation rationalizes the use of Eq. (4.79) for calculating the entropy. The finding is also in harmony with the theoretical characterization of the conformational fluctuation of protein in solution formulated in the preceding section in the chapter. That is, the structural fluctuation takes place on the free energy surface which is quadratic with respect to the displacement of a biomolecule in water is *Gaussian*.

Entropy of Water Around a Biomolecule: There is another type of excess entropy that plays a crucial role in a biomolecular process. It is concerned with so-called *molecular-recognition* process in which a substrate molecule in an enzymatic reaction, or a drug compound, is bound at the active site of a protein. In those cases, the binding affinity is determined essentially by two factors: the interaction between amino acids forming the cavity and the ligand molecule, and change of *entropy* concerning the biomolecule and solution species including water and ligand. Then, why *entropy* of water is important? Let us imagine the process of ligand binding by a protein. When a ligand molecule is bound at an active site, two processes of the entropy change are associated with. Firstly, the ligand molecule will lose entropy during the binding process, because the molecule is confined in the cavity, and loses the degrees of freedom. On the other hand, some water molecules should be disposed from the cavity. Those water molecules will gain entropy, because they acquire the degrees of freedom to move around inside a cell. Efforts have been devoted to evaluate the entropy of water around a biomolecule by means of the molecular-dynamics simulation (Gilson et al. 1997; Lazaridis 1998a, b; Sindhikara and Hirata 2013). It is however a highly nontrivial problem for the method to evaluate the entropy of water molecules, especially, those confined in a cavity such as an active site of an enzyme. Those water molecules are exchanging their position with those residing outside the cavity, and the exchange rate is usually very slow. So, it is virtually impossible to sample all configurations of water molecules around a cavity of a biomolecule by the simulation.

In the present section, we derive equations to evaluate the entropy of a biomolecule in water and that of water around a biomolecule based on the Kim–Hirata theory. We employ the Gibbs formula Eq. (4.80) below to derive the entropy (Chandler 1987; Hirata 2020).

$$S = -k_{\mathrm{B}} \int w(\Delta x) \log w(\Delta x) \mathrm{d}x, \tag{4.80}$$

where $\Delta x (= x - \langle x \rangle)$ represents the fluctuation of the event x from its average value, $\langle x \rangle$, and $w(\Delta x)$ denotes the probability distribution of the event to take a value Δx. So, the crucial step for formulating the entropy is to find the *probability distribution* for the two types of fluctuation, the conformational fluctuation of a solute and the density fluctuation of solvent around the solute.

The *excess* entropy related to a biomolecule in water has its origin in the two kinds of fluctuations: the structural fluctuation of protein and the density fluctuation of solvent.

$$S_{\mathrm{excess}} = S_{\mathrm{conf}} + S_{\mathrm{solv}} \tag{4.81}$$

In the equation, S_{conf} and S_{solv} represent the entropies associated with the structural fluctuation and the density fluctuation, respectively.

It is noted that S_{conf} involves a contribution from solvent around solute. However, the contribution should not be confused with S_{solv}. S_{solv} is the entropy originated from the density fluctuation of solvent around solute with a *fixed* structure of protein. On the other hand, the solvent contribution involved in S_{conf} is related to the density fluctuation of solvent *conjugated* with the conformational fluctuation of protein. Therefore, it is a cross term of the structural fluctuation of protein and density fluctuation of water.

In the following, we formulate the theory to evaluate the two kinds of entropies based on the Kim–Hirata theory explained in the previous sections in the chapter.

4.3.2 Entropy Associated with Conformational Fluctuation of a Solute Molecule

The entropy associated with the conformational fluctuation, defined in terms of the Gibbs formula (Eq. 4.80) is,

$$S_{\text{conf}} = -k_{\text{B}} \int w_{\text{conf}}(\{\Delta\mathbf{R}\}) \log w_{\text{conf}}(\{\Delta\mathbf{R}\}) \prod_{\alpha} \mathrm{d}\Delta\mathbf{R}_{\alpha} \tag{4.82}$$

where $\Delta\mathbf{R}_{\alpha}(=\mathbf{R}_{\alpha} - \langle\mathbf{R}_{\alpha}\rangle)$ represents the positional fluctuation of atom α in protein, and $w(\{\Delta\mathbf{R}\})$ is the probability distribution of the conformational fluctuation defined by Eq. (4.61). Substituting Eq. (4.61) into Eq. (4.82) leads,

$$\begin{aligned} S_{\text{conf}} = &-k_{\text{B}} \int \cdots \int \sqrt{\frac{\|K\|}{(2\pi)^{3N}}} \exp\left[-\frac{1}{2}\sum_{\alpha}\sum_{\beta} K_{\alpha\beta}\Delta\mathbf{R}_{\alpha}\Delta\mathbf{R}_{\beta}\right] \\ &\times \log\sqrt{\frac{\|K\|}{(2\pi)^{3N}}} \exp\left[-\frac{1}{2}\sum_{\alpha}\sum_{\beta} K_{\alpha\beta}\Delta\mathbf{R}_{\alpha}\Delta\mathbf{R}_{\beta}\right] \prod_{\alpha} \mathrm{d}\Delta\mathbf{R}_{\alpha}, \end{aligned} \tag{4.83}$$

where $\|K\|$ is the determinant of the matrix $\{K_{\alpha\beta}\}$ that is the inverse of the variance–covariance matrix, $\{L_{\alpha\beta}\} \equiv \langle\Delta\mathbf{R}_{\alpha}\Delta\mathbf{R}_{\beta}\rangle$.

The integral in Eq. (4.83) can be split into two terms as follows.

$$\begin{aligned} S_{\text{conf}} = &-k_{\text{B}} \int \cdots \int \sqrt{\frac{\|K\|}{(2\pi)^{3N}}} \exp\left[-\frac{1}{2}\sum_{\alpha}\sum_{\beta} K_{\alpha\beta}\Delta\mathbf{R}_{\alpha}\Delta\mathbf{R}_{\beta}\right] \\ &\times \log\left\{\sqrt{\frac{\|K\|}{(2\pi)^{3N}}} \exp\left[-\frac{1}{2}\sum_{\alpha}\sum_{\beta} K_{\alpha\beta}\Delta\mathbf{R}_{\alpha}\Delta\mathbf{R}_{\beta}\right]\right\} \prod_{\alpha} \mathrm{d}\Delta\mathbf{R}_{\alpha} \\ = &-k_{\text{B}} \int \cdots \int \sqrt{\frac{\|K\|}{(2\pi)^{3N}}} \exp\left[-\frac{1}{2}\sum_{\alpha}\sum_{\beta} K_{\alpha\beta}\Delta\mathbf{R}_{\alpha}\Delta\mathbf{R}_{\beta}\right] \\ &\times \left\{\log\sqrt{\frac{\|K\|}{(2\pi)^{3N}}} - \frac{1}{2}\sum_{\alpha}\sum_{\beta} K_{\alpha\beta}\Delta\mathbf{R}_{\alpha}\Delta\mathbf{R}_{\beta}\right\} \prod_{\alpha} \mathrm{d}\Delta\mathbf{R}_{\alpha} \\ = &-k_{\text{B}} \int \cdots \int \sqrt{\frac{\|K\|}{(2\pi)^{3N}}} \exp\left[-\frac{1}{2}\sum_{\alpha}\sum_{\beta} K_{\alpha\beta}\Delta\mathbf{R}_{\alpha}\Delta\mathbf{R}_{\beta}\right] \\ &\times \log\sqrt{\frac{\|K\|}{(2\pi)^{3N}}} \prod_{\alpha} \mathrm{d}\Delta\mathbf{R}_{\alpha} \\ &+ k_{\text{B}} \int \cdots \int \sqrt{\frac{\|K\|}{(2\pi)^{3N}}} \exp\left[-\frac{1}{2}\sum_{\alpha}\sum_{\beta} K_{\alpha\beta}\Delta\mathbf{R}_{\alpha}\Delta\mathbf{R}_{\beta}\right] \\ &\times \left\{\frac{1}{2}\sum_{\alpha}\sum_{\beta} K_{\alpha\beta}\Delta\mathbf{R}_{\alpha}\Delta\mathbf{R}_{\beta}\right\} \prod_{\alpha} \mathrm{d}\Delta\mathbf{R}_{\alpha} \end{aligned}$$

$$= S_1 + S_2 \tag{4.84}$$

The Gaussian integral of the first term can be readily carried out to give

$$S_1 = -k_B \int \cdots \int \sqrt{\frac{\|K\|}{(2\pi)^3}} \exp\left[-\frac{1}{2}\sum_\alpha \sum_\beta K_{\alpha\beta} \Delta \mathbf{R}_\alpha \Delta \mathbf{R}_\beta\right] \log \sqrt{\frac{\|K\|}{(2\pi)^{3N}}} \prod_\alpha \mathrm{d}\mathbf{R}_\alpha$$

$$= \frac{1}{2} k_B \log\left(\frac{(2\pi)^{3N}}{\|K\|}\right) \tag{4.85}$$

In order to carry out the integral in the second term of Eq. (4.85), we introduce a parameter x.

$$Y = k_B \int \cdots \int \sqrt{\frac{\|K\|}{(2\pi)^{3N}}} \exp\left[-x\frac{1}{2}\sum_\alpha \sum_\beta K_{\alpha\beta} \Delta \mathbf{R}_\alpha \Delta \mathbf{R}_\beta\right] \prod_\alpha \mathrm{d}\mathbf{R}_\alpha. \tag{4.86}$$

Then, the second term in Eq. (4.85) can be then written as

$$S_2 = -\left.\frac{\mathrm{d}Y}{\mathrm{d}x}\right|_{x=1} \tag{4.87}$$

The integral in Eq. (4.86) can be carried out readily to give

$$Y = k_B \int \sqrt{\frac{\|K\|}{(2\pi)^{3N}}} \exp\left[-\frac{1}{2}\sum_\alpha \sum_\beta x K_{\alpha\beta} \Delta \mathbf{R}_\alpha \Delta \mathbf{R}_\beta\right] \prod_\alpha \mathrm{d}\Delta\mathbf{R}_\alpha$$

$$= k_B \sqrt{\frac{\|K\|}{(2\pi)^{3N}}} \sqrt{\frac{(2\pi)^{3N}}{x^{3N} A}} = k_B x^{-3N/2} \tag{4.88}$$

Therefore,

$$S_2 = -\left.\frac{\mathrm{d}Y}{\mathrm{d}x}\right|_{x=1} = \frac{3N}{2} k_B \tag{4.89}$$

We find finally for the conformational entropy of protein as

$$S_{\mathrm{conf}} = S_1 + S_2$$

$$= \frac{1}{2} k_B \log\left(\frac{(2\pi)^{3N}}{\|K\|}\right) + \frac{3N}{2} k_B. \tag{4.90}$$

In the equation, $\|K\|$ is the determinant of the matrix $\{K_{\alpha\beta}\}$, the element of which is the second derivative of the free energy, defined by Eq. (4.59), with respect to the coordinates of atom α and β. It is in turn the inverse of the determinant of the variance–covariance matrix as is defined by Eqs. (4.17) and (4.54). So, the expression is in accord with our intuition, that is, *greater the variance, larger the conformational entropy*.

4.3.3 Entropy Associated with Solvent-Density Fluctuation Around a Biomolecule in Solution

In order to complete the mathematical isomorphism with a harmonic oscillator, we define the free energy of solvent by,

$$F_{\text{solv}} \equiv \int\int \mu_{\text{solv}}(\mathbf{r}, \mathbf{r}')d\mathbf{r}d\mathbf{r}' \tag{4.91}$$

where the *local free energy* $\mu_{\text{solv}}(\mathbf{r}, \mathbf{r}')$ at a pair of positions $\mathbf{r}$ and $\mathbf{r}'$ is defined by

$$\mu_{\text{solv}}(\mathbf{r}, \mathbf{r}') \equiv \frac{1}{2}k_BT\sum_a\sum_b \delta\rho_a(\mathbf{r})\kappa_{ab}^{\text{solv}}(\mathbf{r}, \mathbf{r}')\delta\rho_b(\mathbf{r}') \tag{4.92}$$

Then, $\kappa_{ab}^{\text{solv}}(\mathbf{r}, \mathbf{r}')$ is identified as the second derivative of the free energy F_{solv} with respect to the density fluctuation at fixed positions $\mathbf{r}$ and $\mathbf{r}'$,

$$\kappa_{ab}^{\text{solv}}(\mathbf{r}, \mathbf{r}') = \frac{\delta^2(F_{\text{solv}}/k_BT)}{\delta\rho_a(\mathbf{r})\delta\rho_b(\mathbf{r}')}. \tag{4.93}$$

The identification is similar to the density-functional theory of liquids, except for the kinetic factor defined by (4.72). The kinetic factor is included there, because the momentum density or flux also contributes to change of the free energy.

The physics of the equation can be interpreted in an analogy to the harmonic oscillator as follows. The *force* to restore the *equilibrium density* of solvent-atom a at position $\mathbf{r}$ is proportional to the density fluctuation of solvent-atom b at position $\mathbf{r}'$, and the *force constant* $\kappa_{ab}^{\text{solv}}(\mathbf{r}, \mathbf{r}')$ of the *restoring force* is the second derivative of the free energy with respect to the density fluctuation of atoms a and b.

The probability distribution of the density fluctuation of solvent atoms a and b at $\mathbf{r}$ and $\mathbf{r}'$, having values $\delta\rho_a(\mathbf{r})$ and $\delta\rho_b(\mathbf{r}')$, respectively, is

$$\begin{aligned} w_{\text{solv}}\left(\{\delta\rho\}; \mathbf{r}, \mathbf{r}'\right) &\propto \exp\left[-\frac{\mu_{\text{solv}}(\mathbf{r}, \mathbf{r}')}{k_BT}\right] \\ &= \sqrt{\frac{\|B\|}{(2\pi)^{3n}}}\exp\left[-\frac{1}{2}\sum_a\sum_b \delta\rho_a(\mathbf{r})\kappa_{ab}^{\text{solv}}(\mathbf{r}, \mathbf{r}')\delta\rho_b(\mathbf{r}')\right] \end{aligned} \tag{4.94}$$

where B is the normalization constant, and it is the determinant of the matrix $\left\{\kappa_{ab}^{\rm solv}(\mathbf{r},\mathbf{r}')\right\}$. The expression is apparently a Gaussian distribution with respect to the density fluctuations at the fixed position $\mathbf{r}$ and $\mathbf{r}'$, the variance of which is $\kappa_{ab}^{\rm solv}(\mathbf{r},\mathbf{r}')^{-1}$.

We define a quantity referred to as the *local solvent entropy* $s_{\rm solv}(\mathbf{r},\mathbf{r}')$ at a pair of positions, $\mathbf{r}$ and $\mathbf{r}'$, as

$$s_{\rm solv}(\mathbf{r},\mathbf{r}') \equiv -k_{\rm B}\int\cdots\int w_{\rm solv}\big(\{\delta\rho\};\mathbf{r},\mathbf{r}'\big)\log w_{\rm solv}\big(\{\delta\rho\};\mathbf{r},\mathbf{r}'\big)\prod_a \mathrm{d}\delta\rho_a, \quad (4.95)$$

where $w_{\rm solv}$ is defined by Eq. (4.94).

It should be noted that the variable of integral is $\delta\rho(\mathbf{r})$, not the coordinate $\mathbf{r}$, since $\mathbf{r}$ is fixed. The density fluctuation here is a temporal fluctuation rather than a spatial fluctuation. The entropy of solvation at position $\mathbf{r}$ around a biomolecule, and of the entire system is defined, respectively, as

$$s_{\rm solv}(\mathbf{r}) \equiv \frac{1}{V}\int s_{\rm solv}(\mathbf{r},\mathbf{r}')\mathrm{d}\mathbf{r}' \quad (4.96)$$

and

$$s_{\rm solv} \equiv \frac{1}{V^2}\int\int s_{\rm solv}(\mathbf{r},\mathbf{r}')\mathrm{d}\mathbf{r}\mathrm{d}\mathbf{r}' \quad (4.97)$$

The local entropy is obtained by substituting the expression of Eq. (4.94) for $w_{\rm solv}$ into Eq. (4.95), and performing the integral with respect to the density fluctuation. The integral may be carried out as follows.

$$\begin{aligned}
s_{\rm solv}(\mathbf{r},\mathbf{r}') &= -k_{\rm B}\int\cdots\int\sqrt{\frac{\|B\|}{(2\pi)^{3n}}}\exp\left[-\frac{1}{2}\sum_a\sum_b \kappa_{ab}\delta\rho_a\delta\rho_b\right] \\
&\quad\times\log\left\{\sqrt{\frac{\|B\|}{(2\pi)^{3n}}}\exp\left[-\frac{1}{2}\sum_a\sum_b \kappa_{ab}\delta\rho_a\delta\rho_b\right]\right\}\prod_a \mathrm{d}\delta\rho_a \\
&= -k_{\rm B}\int\cdots\int\sqrt{\frac{\|B\|}{(2\pi)^{3n}}}\exp\left[-\frac{1}{2}\sum_a\sum_b \kappa_{ab}\delta\rho_a\delta\rho_b\right] \\
&\quad\times\left\{\log\sqrt{\frac{\|B\|}{(2\pi)^{3n}}}-\frac{1}{2}\sum_a\sum_b \kappa_{ab}\delta\rho_a\delta\rho_b\right\}\prod_a \mathrm{d}\delta\rho_a \\
&= -k_{\rm B}\int\cdots\int_a\sqrt{\frac{\|B\|}{(2\pi)^{3n}}}\exp\left[-\frac{1}{2}\sum_a\sum_b \kappa_{ab}\delta\rho_a\delta\rho_b\right]
\end{aligned}$$

$$
\begin{aligned}
&\log\sqrt{\frac{\|B\|}{(2\pi)^{3n}}}\prod_a \mathrm{d}\delta\rho_a \\
&+k_\mathrm{B}\int\cdots\int\sqrt{\frac{\|B\|}{(2\pi)^{3n}}}\exp\left[-\frac{1}{2}\sum_a\sum_b \kappa_{ab}\delta\rho_a\delta\rho_b\right] \\
&\times\left\{\frac{1}{2}\sum_a\sum_b \kappa_{ab}\delta\rho_a\delta\rho_b\right\}\prod_a \mathrm{d}\delta\rho_a \\
&\equiv s^{(1)}_\mathrm{solv}(\mathbf{r},\mathbf{r}')+s^{(2)}_\mathrm{solv}(\mathbf{r},\mathbf{r}')
\end{aligned}
\tag{4.98}
$$

where $s^{(1)}_\mathrm{solv}(\mathbf{r},\mathbf{r}')$ and $s^{(2)}_\mathrm{solv}(\mathbf{r},\mathbf{r}')$ are defined as,

$$
\begin{aligned}
s^{(1)}_\mathrm{solv}(\mathbf{r},\mathbf{r}') &= k_\mathrm{B}\int\cdots\int\sqrt{\frac{\|B\|}{(2\pi)^{3n}}}\exp\left[-\frac{1}{2}\sum_a\sum_b \kappa_{ab}\delta\rho_a\delta\rho_b\right] \\
&\quad\log\sqrt{\frac{\|B\|}{(2\pi)^{3n}}}\prod_a \mathrm{d}\delta\rho_a \\
s^{(2)}_\mathrm{solv}(\mathbf{r},\mathbf{r}') &= k_\mathrm{B}\int\cdots\int\sqrt{\frac{\|B\|}{(2\pi)^{3n}}}\exp\left[-\frac{1}{2}\sum_a\sum_b \kappa_{ab}\delta\rho_a\delta\rho_b\right] \\
&\quad\times\left\{\frac{1}{2}\sum_a\sum_b \kappa_{ab}\delta\rho_a\delta\rho_b\right\}\prod_a \mathrm{d}\delta\rho_a
\end{aligned}
\tag{4.99}
$$

The integrals are carried out as follows, since they are just a Gaussian integral.

$$
\begin{aligned}
s^{(1)}(\mathbf{r},\mathbf{r}') &= -k_\mathrm{B}\int\cdots\int\prod_a \mathrm{d}\delta\rho_a\sqrt{\frac{\|B\|}{(2\pi)^{3}}}\exp\left[-\frac{1}{2}\sum_a\sum_b \kappa_{ab}\delta\rho_a\delta\rho_b\right] \\
&\quad\log\sqrt{\frac{\|B\|}{(2\pi)^{3N}}} \\
&= \frac{1}{2}k_\mathrm{B}\log\left(\frac{(2\pi)^n}{\|B\|}\right)
\end{aligned}
\tag{4.100}
$$

$$
\begin{aligned}
s^{(2)}_\mathrm{solv}(\mathbf{r},\mathbf{r}') &= k_\mathrm{B}\int\cdots\int\prod_a \mathrm{d}\delta\rho_a\sqrt{\frac{\|B\|}{(2\pi)^{3n}}}\exp\left[-\frac{1}{2}\sum_a\sum_b \kappa_{ab}\delta\rho_a\delta\rho_b\right] \\
&\quad\times\left\{\frac{1}{2}\sum_a\sum_b \kappa_{ab}\delta\rho_a\delta\rho_b\right\} \\
&= \frac{3n}{2}k_\mathrm{B}
\end{aligned}
\tag{4.101}
$$

In the equation, $\|B\|$ is the determinant of the matrix $\left\{\kappa_{ab}^{\text{solv}}(\mathbf{r},\mathbf{r}')\right\}$ at the fixed position $\mathbf{r}$ and $\mathbf{r}'$, the element of which is the inverse of the site–site pair correlation function multiplied by the kinetic factor $\{J_{ac}\}$ as defined by Eq. (4.71). From Eqs. (4.98), (4.100) and (4.101), one finds an expression for the local entropy of solvent, $s_{\text{solv}}(\mathbf{r},\mathbf{r}')$, as,

$$s_{\text{solv}}(\mathbf{r},\mathbf{r}') = \frac{1}{2}k_{\text{B}}\log\left(\frac{(2\pi)^n}{\|B\|}\right) + \frac{3n}{2}k_{\text{B}} \tag{4.102}$$

Thus, microscopic equations are derived for the two types of entropy in biomolecular solutions: the conformational entropy and the entropy of solvent around a solute. The conformational entropy is concerned with the *inverse* of the *force constant, or Hessian,* of the restoring force of structural fluctuation of a biomolecule in solution. The force constant turns out to be an inverse of the variance–covariance matrix of the structural fluctuation. The local entropy of solvent around a biomolecule is related to the inverse of the *force constant*, or the second-derivative, of the free energy of solvent with respect to the density fluctuation of solvent at a pair of positions. It is essentially the inverse of the density pair-correlation function which is a measure of the variance of the density fluctuation.

References

Anfinsen CB (1973) Principle that govern the folding of protein chain. Science 181:223–230

Bryngelson JD, Onuchic JN, Socci ND, Wolynes PG (1995) Funnels, pathways, and the energy landscape of protein folding: a synthesis. Proteins 21:167–195

Chandler D (1987) Introduction to modern statistical mechanics. Oxford, New York

Chong S-H, Ham S (2011) Configurational entropy of protein: a combined approach based on molecular simulation and integral-equation theory of liquids. Chem Phys Lett 504(46):225–229

Chong S-H, Ham S (2013) Conformational entropy of intrinsically disordered protein. J Phys Chem B 117:5503–5509

Chong S-H, Ham S (2016) New computational approach for external entropy in protein-protein binding. J Chem Theory Comput 12(6):2509–2516

Gilson MK, Given JA, Bush BL, McCammon JA (1997) The statistical-thermodynamics basis for computation of binding affinity. Biophys J 72:1057–1069

Go N, Go M, Scheraga HA (1968) Molecular theory of the helix-coil transition in polyamino acids, I. Formulation. Proc Natl Acad Sci USA 59:1030–1037

Go M, Go N, Scheraga HA (1970) J Chem Phys 52:2060

Hansen JP, McDonald IR (1986) Theory of simple liquids. Academic Press, London

Hirata F (2020) Exploring life phenomena with statistical mechanics of molecular liquids. CRC Press, New York

Karplus M, Kushick JN (1981) Method for estimating the configurational entropy of macromolecules. Macromolecules 14:325–332

Kataoka M, Nishi I, Fujisawa T, Ueki T, Tokunaga F, Goto Y (1995) Structural characterization of the molten globule and native states of apomyoglobin by solution X-ray scattering. J Mol Biol 249:215–228

Kim B, Hirata F (2013) Structural fluctuation of protein in water around its native state: a new statistical mechanics formulation. J Chem Phys 138:054108–054111

Lamm G, Szabo A (1986) Langevin mode of macromolecules. J Chem Phys 85:7334
Lazaridis T (1998a) Inhomogeneous fluid approach to solvation thermodynamics. 1. Theory. J Phys Chem 102:3531–3541
Lazaridis T (1998b) Inhomogeneous fluid approach to solvation thermodynamics. 2. Applications to simple fluids. J Phys Chem 102:3542–3550
Lazaridis T, Karplus M (1996) Orientational correlations and entropy in liquid water. J Chem Phys 105:4294–4316
Levinthal C (1969) How to fold graciously. Abstract of the talk. In: Mossbauer spectroscopy in biological systems: proceedings of a meeting. Allerton House, Monticello, Illinois
Levy RM, Karplus M, Kushick JN, Perahia D (1981) Evaluation of the configurational entropy for proteins: application to molecular dynamics simulations of an α-helix. Macromolecules 14:325–332
McQuarrie DA (1976) Statistical mechanics. Harper and Row, New York
Prigogine I, Defay R (1966) Thermodynamique chimique (Japanese translation). Misuzu Publishing Co.
Rowlinson JS (1969) Liquids and liquid mixtures. Butterworths, London
Schlitter J (1993) Estimation of absolute and relative entropies of macromolecules using the covariance matrix. Chem Phys Lett 215:617–621
Sindhikara DJ, Hirata F (2013) Analysis of biomolecular solvation sites by 3D-RISM theory. J Phys Chem B 117:6718–6723
Wang MC, Uhlenbeck GE (1945) On the theory of Brownian motion II. Rev Mod Phys 17:323–342

Chapter 5
Time Evolution of Structural Fluctuation of a Biomolecule in Water

In the preceding chapter, we have derived the generalized Langevin equations that describe the structural fluctuation of a biomolecule and the density fluctuation of water, which are coupled together. The most essential finding in the theory is that the free energy surface to produce the restoring force is *quadratic* with respect to the structural fluctuation or the displacement, which indicates that the structural fluctuation of a biomolecule in water is *harmonic*, and that the probability distribution of the fluctuation is *Gaussian.* It is a common strategy in the theoretical physics and chemistry to apply the *harmonic analysis* to such a process that is taking place on a quadratic surface. By performing the harmonic analysis, or solving an eigen value problem, one may decompose a multi-dimensional free-energy-surface into independent quadratic surfaces with different curvatures, which are hierarchically ordered.

The present chapter is devoted to develop a computational model for the biophysics and chemistry to carry out the harmonic analysis, which is tractable but maintains essential physics of the fluctuation. Two methods will be explored, respectively, in Sects. 5.1 and 5.2, respectively.

First method presented in Sect. 5.1 is almost identical to that proposed by Lamm and Szabo in 1986 except for the model and for the starting equation (Lamm and Szabo 1986). The authors have started from the classical Langevin equation applied to a *harmonic-oscillator* model of protein, proposed much earlier by Wang and Uhlenbeck, in which the atomic motion evolves on a *potential-energy* surface produced by the *direct interactions* among atoms (Wang and Uhlenbeck 1945). On the other hand, the model presented here is based on the *generalized* Langevin equations of the structural fluctuation of a protein in solution, derived in the preceding chapter. In the dynamics, each atom is driven by a force derived from the *free-energy* surface, that includes not only the *direct interaction* among atoms but also the solvent induced force (Kim and Hirata 2013).

The second treatment presented in Sect. 5.2 is more or less similar to the ordinary molecular dynamics (MD) simulation in the sense that it carries out a numerical

F. Hirata, *Molecular Theory of Fluctuation in Life Phenomena*,
https://doi.org/10.1007/978-981-96-3364-7_5

integration of the GLE equation (Eq. 5.25) with the Verlet-type algorithm, although there are two essential differences between the two methods. The first of those is the force to drive atoms in a molecule. The *first derivative* of the *potential energy* with respect to the coordinates of atoms is used as the driving force in MD, while the *second derivative* of the *free energy* is used to calculate the force in the present method.

The other difference is related to the time step. In the usual MD simulation, the time step should be small enough, ~ 10^{-15} s or so, to maintain the dynamics in healthy condition. On the other hand, the dynamics explored in the chapter enables to decouple the time evolution into modes with different frequencies in hierarchical order, which in turn allow us to perform multiscale dynamics of a biomolecule in water.

5.1 Generalized Langevin Mode Analysis of a Biomolecule in Water

The structural fluctuation of a biomolecule in solution is defined by,

$$\Delta \mathbf{R}_\alpha(t) \equiv \mathbf{R}_\alpha(t) - \langle \mathbf{R}_\alpha \rangle \tag{5.1}$$

where $\mathbf{R}_\alpha(t)$ is the current position of atom α, and $\langle \mathbf{R}_\alpha \rangle$ defines the equilibrium position of the atom α. The generalized Langevin equations of the structural fluctuation are written as,

$$\frac{d\Delta \mathbf{R}_\alpha(t)}{dt} = \Delta \mathbf{V}_\alpha(t) \tag{5.2}$$

$$\begin{aligned} \frac{d\Delta \mathbf{V}_\alpha(t)}{dt} &= -\frac{k_B T}{M_\alpha} \sum_\beta \left(\mathbf{L}^{-1}\right)_{\alpha\beta} \Delta \mathbf{R}_\alpha(t) \\ &\quad - \int_0^t ds \sum_\beta \frac{1}{M_\alpha} \Gamma_{\alpha\beta}(t-s) \Delta \mathbf{V}_\beta(s) + \mathbf{W}_\alpha(t) \end{aligned} \tag{5.3}$$

where $\mathbf{L}$ is the variance–covariance matrix of structural fluctuation, defined by

$$L_{\alpha\beta} \equiv \langle \Delta \mathbf{R} \Delta \mathbf{R} \rangle_{\alpha\beta}. \tag{5.4}$$

$\Gamma_{\alpha\beta}(t)$ is the friction kernel, and $\mathbf{W}_\alpha(t)$ represents the random force. The variance covariance matrix is related to the second derivative of the free energy surface $F(\{\mathbf{R}\})$ with respect to the atomic coordinates of protein as,

$$k_{\mathrm{B}}T(\mathbf{L}^{-1})_{\alpha\beta} = \frac{\partial^2 F(\{\mathbf{R}\})}{\partial \mathbf{R}_\alpha \partial \mathbf{R}_\beta}. \tag{5.5}$$

The free energy surface is defined as the sum of the conformational energy $U(\{\mathbf{R}\})$ of the solute and the solvation free-energy $\Delta\mu(\{\mathbf{R}\})$ as follows,

$$F(\{\mathbf{R}\}) = U(\{\mathbf{R}\}) + \Delta\mu(\{\mathbf{R}\}). \tag{5.6}$$

Multiplying Eqs. (5.2) and (5.3) by $\Delta\mathbf{R}_\gamma(0)$, and taking the ensemble average, one finds

$$\frac{\mathrm{d}\langle\Delta\mathbf{R}_\alpha(t)\Delta\mathbf{R}_\gamma(0)\rangle}{\mathrm{d}t} = \langle\Delta\mathbf{V}_\alpha(t)\Delta\mathbf{R}_\gamma(0)\rangle, \tag{5.7}$$

$$\begin{aligned}\frac{\mathrm{d}\langle\Delta\mathbf{V}_\alpha(t)\Delta\mathbf{R}_\gamma(0)\rangle}{\mathrm{d}t} = &-\frac{k_{\mathrm{B}}T}{M_\alpha}\sum_\beta \left(\mathbf{L}^{-1}\right)_{\alpha\beta}\langle\Delta\mathbf{R}_\alpha(t)\Delta\mathbf{R}_\gamma(0)\rangle \\ &-\int_0^t \mathrm{d}s \sum_\beta \frac{1}{M_\alpha}\Gamma_{\alpha\beta}(t-s)\langle\Delta\mathbf{V}_\beta(s)\Delta\mathbf{R}_\gamma(0)\rangle \\ &+\langle\Delta\mathbf{W}_\alpha(s)\Delta\mathbf{R}_\gamma(0)\rangle.\end{aligned} \tag{5.8}$$

The last term in the second equation vanishes by definition, because $\Delta\mathbf{R}_\gamma(0)$ is one of the dynamic variables, which is orthogonal to the random force $\Delta\mathbf{W}_\alpha$, that is,

$$\langle\Delta\mathbf{W}_\alpha(s)\Delta\mathbf{R}_\gamma(0)\rangle = 0. \tag{5.9}$$

We further assume that the friction coefficient is local in time, or

$$\Gamma_{\alpha\beta}(t-s) = \zeta_{\alpha\beta}\delta(t-s). \tag{5.10}$$

Taking those into account, the second equation becomes,

$$\begin{aligned}\frac{\mathrm{d}\langle\Delta\mathbf{V}_\alpha(t)\Delta\mathbf{R}_\gamma(0)\rangle}{\mathrm{d}t} = &-\sum_\beta K_{\alpha\beta}\langle\Delta\mathbf{R}_\beta(t)\Delta\mathbf{R}_\gamma(0)\rangle \\ &-\sum_\beta \zeta_{\alpha\beta}\langle\Delta\mathbf{V}_\beta(t)\Delta\mathbf{R}_\gamma(0)\rangle,\end{aligned} \tag{5.11}$$

where $K_{\alpha\beta}$ and $\zeta_{\alpha\beta}$ are defined by

$$K_{\alpha\beta} \equiv \frac{k_{\mathrm{B}}T}{M_\alpha}\left(\mathbf{L}^{-1}\right)_{\alpha\beta}, \quad \zeta_{\alpha\beta} \equiv \frac{\Gamma_{\alpha\beta}}{M_\alpha}. \tag{5.12}$$

Equations (5.7) and (5.11) are written in a matrix form as

$$\frac{\mathrm{d}}{\mathrm{d}t}\begin{pmatrix}\mathbf{C}(t)\\ \dot{\mathbf{C}}(t)\end{pmatrix}=\mathbf{A}\begin{pmatrix}\mathbf{C}(t)\\ \dot{\mathbf{C}}(t)\end{pmatrix}, \tag{5.13}$$

where $C(t)$ and $\dot{\mathbf{C}}(t)$ are defined by

$$C_{\alpha\gamma}(t)\equiv\left\langle\Delta\mathbf{R}_{\alpha}(t)\Delta\mathbf{R}_{\gamma}(0)\right\rangle,\quad \dot{C}_{\alpha\gamma}(t)\equiv\left\langle\Delta\mathbf{V}_{\alpha}(t)\Delta\mathbf{R}_{\gamma}(0)\right\rangle, \tag{5.14}$$

and the matrix $\mathbf{A}$ is defined by the following equation.

$$\mathbf{A}=\begin{pmatrix}\mathbf{0} & \mathbf{1}\\ -\mathbf{K} & -\zeta\end{pmatrix}. \tag{5.15}$$

We refer to the matrix $\mathbf{A}$ as "generalized force constant (GFC) matrix". As it is clear from the definition, GFC carries not only the information concerning the second derivative of the *free energy* surface, but also the influence due to the *friction* from solvent. So, the generalization of the force constant compared to the usual force constant of a molecule situated in *vacuum* is double fold: inclusion of the solvation free energy and the friction from solvent. It should be noted, however, that the basic functional form of the free energy surface on which the structure fluctuates is quadratic. The kinetic energy is also *quadratic* function of the momentum. Therefore, the harmonic analysis analogous to the normal mode analysis is valid for the generalized *Hessian* (Eq. 5.15) in the phase space.

If the force constant matrix $\mathbf{K}$ is replaced by the second derivative of the *potential energy* among atoms in the biomolecule, and the friction-coefficient matrix ζ by the Oseen tensor that takes account for the hydrodynamic force, the equation reduces to that of the Langevin mode analysis (LMA) derived by Lamm and Szabo (1986).

Free Energy Surface and Force Constant Matrix: The free energy surface of the solute is defined by Eq. (5.6) as a sum of the conformational energy of solute and the solvation free energy. It is a routine work in the molecular mechanics or dynamics to calculate the conformational energy $U(\{\mathbf{R}\})$ and its first and the second derivatives with respect to the coordinates $\{\mathbf{R}\}$ for a particular conformation of solute.

On the other hand, there is no such a simple method to calculate the solvation free-energy, or $\Delta\mu(\{\mathbf{R}\})$, by means of the molecular mechanics and dynamics. It is because the method should sample all the configurational space of solvent for a *single* conformation of the solute, which by itself takes the time equivalent to the *Ergodic limit*. It is even more difficult and virtually hopeless to calculate the second derivative, or the force constant, of the free energy surface by means of the MD methods. Fortunately, such quantities as the solvation free energy and its derivatives can be calculated based on an alternative method of the statistical mechanics, or the RISM/3D-RISM theory that has been reviewed in Chap. 2.

Calculation of the Hessian Matrix: The calculation of the second derivative of $U(\{\mathbf{R}\})$ has been a routine task of molecular mechanics, provided that the potential-energy function for protein is given. However, that for the solvation free energy

$\Delta\mu(\{\mathbf{R}\})$ is a non-trivial problem by any means. First of all, it is impossible for the molecular simulation, because the molecular simulation does not give the solvation free energy as an explicit function of the atomic coordinates $\{\mathbf{R}\}$ of a biomolecule in water. The method may be able to calculate the potential of mean force projected along a very limited space of the *reaction coordinates* by means of the umbrella sampling. However, it cannot calculate the free energy surface for the entire conformational-space spanned by $\{\mathbf{R}\}$. So, there is no hope for the method to evaluate the first and second derivatives of the solvation free energy.

On the other hand, the calculation of $\Delta\mu(\{\mathbf{R}\})$ as a function of the atomic coordinates $\{\mathbf{R}\}$ of a biomolecule is a routine task for the methods based on the statistical mechanics of liquids represented by the RISM/3D-RISM theory (Kovalenko and Hirata 1998, 1999).

The method to calculate the first derivative, or the "solvent-mediated force", has been derived by Yoshida and Hirata as,

$$\frac{\partial}{\partial \mathbf{R}_\alpha}\Delta\mu(\{\mathbf{R}\}) = \sum_i \rho_i \int \frac{\partial u_i^{uv}(\mathbf{r};\{\mathbf{R}\})}{\partial \mathbf{R}_\alpha} g_i^{uv}(\mathbf{r};\{\mathbf{R}\})d\mathbf{r}, \tag{5.16}$$

where ρ_i is the number density of solvent atom i, $u_i^{uv}(\mathbf{r};\{\mathbf{R}\})$ is the interaction between the solute molecule and solvent atom i residing at the position $\mathbf{r}$, and $g_i^{uv}(\mathbf{r};\{\mathbf{R}\})$ is the spatial distribution of the solvent atom i around the solute (Yoshida and Hirata 2006). The equation has been first applied by Miyata and Hirata to the molecular-dynamics simulation combined with the RISM/3D-RISM method to calculate the *solvent-induced force* acting on an atom of a solute molecule in water (Miyata and Hirata 2007). The method has been implemented in the MD-simulation software, AMBER, by Kovalenko and Omelyan, to further accelerate the simulation of a biomolecule in water (Omelyan and Kovalenko 2015).

In order to calculate the Hessian matrix, or Eq. (5.5), we have to carry out the derivative of Eq. (5.16) with respect to the atomic coordinate of solute, that is,

$$\begin{aligned}\frac{\partial^2}{\partial \mathbf{R}_\alpha \partial \mathbf{R}_\beta}\Delta\mu(\{\mathbf{R}\}) &= \sum_i \rho_i \int \frac{\partial^2 u_i^{uv}(\mathbf{r};\{\mathbf{R}\})}{\partial \mathbf{R}_\alpha \partial \mathbf{R}_\beta} g_i^{uv}(\mathbf{r};\{\mathbf{R}\})d\mathbf{r} \\ &+ \sum_i \rho_i \int \frac{\partial u_i^{uv}(\mathbf{r};\{\mathbf{R}\})}{\partial \mathbf{R}_\alpha} \frac{\partial}{\partial \mathbf{R}_\beta} g_i^{uv}(\mathbf{r};\{\mathbf{R}\})d\mathbf{r}.\end{aligned} \tag{5.17}$$

It is a rather simple task to calculate the first term which involves the second derivative of the interaction energy between solute and solvent with respect to the atomic coordinates of the solute molecule.

On the other hand, it is a non-trivial problem to calculate the second term, because it involves the derivative of the spatial correlation function $g_i^{uv}(\mathbf{r};\{\mathbf{R}\})$. Fortunately, the recipe to calculate such derivatives of the spatial distribution function has been already proposed by Yu and Karplus few decades ago (Yu and Karplus 1988). The method is closely related to the numerical solution of the RISM/3D-RISM equation to calculate $g_i^{uv}(\mathbf{r};\{\mathbf{R}\})$.

Although the method is common to any closure relation to solve the equation, we just present the procedure corresponding to the Kovalenko–Hirata closure (Kovalenko and Hirata 1999). Then, the RISM/3D-RISM equation consists of the two equations, which are written as

$$h_j^{uv}(\mathbf{r};\{\mathbf{R}\}) = \sum \int \chi_{jl}^{vv}\left(\left|\mathbf{r}-\mathbf{r}'\right|;\{\mathbf{R}\}\right) c_l^{uv}\left(\mathbf{r}';\{\mathbf{R}\}\right) d\mathbf{r}' \equiv \chi_{jl}^{vv} * c_l^{uv}(\mathbf{r};\{\mathbf{R}\}) \tag{5.18}$$

$$h_j^{uv}(\mathbf{r};\{\mathbf{R}\}) = \begin{cases} \exp\left[d_j^{uv}(\mathbf{r};\{\mathbf{R}\})\right] - 1 & \text{for } d_j^{uv}(\mathbf{r};\{\mathbf{R}\}) \le 0 \\ d_j^{uv}(\mathbf{r};\{\mathbf{R}\}) & \text{for } d_j^{uv}(\mathbf{r};\{\mathbf{R}\}) > 0 \end{cases} \tag{5.19}$$

$$d_j^{uv}(\mathbf{r}) \equiv -u_j^{uv}(\mathbf{r})/k_B T + h_j^{uv}(\mathbf{r}) - c_j^{uv}(\mathbf{r}) \tag{5.20}$$

where $h_j^{uv}(\mathbf{r}) \equiv g_j^{uv}(\mathbf{r}) - 1$. If one interprets the equation in terms of the non-linear response theory, χ_{lj}^{vv} is the site–site pair correlation function of solvent that plays the susceptibility or response function to the perturbation $c_j^{uv}(\mathbf{r})$ from the solute molecule (see Chap. 2). The derivative of the correlation functions with respect to the atomic coordinate of protein can be written as

$$\frac{\partial h_j^{uv}(\mathbf{r};\{\mathbf{R}\})}{\partial \mathbf{R}_\alpha} = \sum_j \chi_{jl}^{vv} * \frac{\partial c_j^{uv}(\mathbf{r};\{\mathbf{R}\})}{\partial \mathbf{R}_\alpha} \tag{5.21}$$

$$\frac{\partial h_j^{uv}(\mathbf{r};\{\mathbf{R}\})}{\partial \mathbf{R}_\alpha} = \begin{cases} \left(-\frac{1}{k_B T}\frac{\partial u_j^{uv}(\mathbf{r};\{\mathbf{R}\})}{\partial \mathbf{R}_\alpha} + \frac{\partial t_j^{uv}(\mathbf{r};\{\mathbf{R}\})}{\partial \mathbf{R}_\alpha}\right) \exp\left(-u_j^{uv}(\mathbf{r};\{\mathbf{R}\})/k_B T + t_j^{uv}(\mathbf{r};\{\mathbf{R}\})\right) & \text{for } -u_j^{uv}(\mathbf{r};\{\mathbf{R}\})/k_B T + t_j^{uv}(\mathbf{r};\{\mathbf{R}\}) \le 0 \\ -\frac{1}{k_B T}\frac{\partial u_j^{uv}(\mathbf{r};\{\mathbf{R}\})}{\partial \mathbf{R}_\alpha} + \frac{\partial t_j^{uv}(\mathbf{r};\{\mathbf{R}\})}{\partial \mathbf{R}_\alpha} & \text{for } -u_j^{uv}(\mathbf{r};\{\mathbf{R}\})/k_B T + t_j^{uv}(\mathbf{r};\{\mathbf{R}\}) > 0 \end{cases} \tag{5.22}$$

The derivatives can be calculated as one of the variables along the course of iteration to find the solutions for the correlation functions themselves.

Diagonalization of the Hessian Matrix: It is a rather standard task for a computational science to diagonalize a matrix to find the eigen value and vector. There is no theoretical difficulty left in principle. However, a non-trivial computational problem is left unsolved, which is the *size* of the Hessian matrix. If one tries to apply the method to a real protein in water, the number of atoms may become ~ 10^4 or more. The number of elements of Hessian for such a protein amounts to a square of the number of atoms, that is, ~ 10^8. So, it may be a challenge even for the peta-scale super computer such as "Fugaku". Nevertheless, it is a challenge that is worthwhile to be tried, because it may solve many questions in life phenomena including those in medicine.

Friction Coefficient: In order to solve the GLMA Eqs. (5.13) with (5.14) and (5.15), and to find the relaxation dynamics of a biomolecule in water, not just the *steady-state* fluctuation, one needs to provide another information, that is, the friction coefficient ζ.

Let us imagine solvent environment of protein. Some atoms located at internal regions of protein will not be exposed to solvent at all. Such atoms may not feel any *friction* from solvent. On the other hand, the protein atoms residing at the surface may be exposed almost fully to solvent. Such atoms will feel as much friction as in the bulk solvent. And many atoms may be in between the two cases. Of course, the proportion of exposure of solute atoms to solvent will depend very much on the conformation of protein: most of atoms are exposed to solvent in the denatured state of protein, while many atoms are protected from solvent in the native conformation. So, the question is how to express such inhomogeneous environment of solvent around protein. A practical solution to the problem has been proposed earlier by Hirata and Kim (2016). The solution is to express the friction coefficient $\zeta_{\alpha\beta}$ exerted on protein atom, α, with $\zeta_{\alpha\beta} = \delta_{\alpha\beta}\zeta_\alpha$ where $\delta_{\alpha\beta}$ is the Kronecker delta function. After all, the friction coefficient is defined by

$$\zeta_{\alpha\alpha} = f_\alpha \zeta_{\alpha\alpha,\text{bulk}} \tag{5.23}$$

where $\zeta_{\alpha\alpha,\text{bulk}}$ is the friction exerted on atom α in *bulk* solvent, and f_α is the *accessible surface area* of atom α, defined by,

$$f_\alpha = \frac{g_w(\sigma;\,\text{protein})}{g_w(\sigma;\,\text{bulk})}. \tag{5.24}$$

In the equation, $g_w(\sigma;\,\text{bulk})$ is the pair correlation function of solvent in bulk at the contact separation between the atom and solvent molecules, and $g_w(\sigma;\,\text{protein})$ is the counter-part for the atom in protein. The equation would meet the requirement for f_α to represent the inhomogeneity of solvent environment around protein mentioned above. If the atom resides at the surface of protein, then a large portion of surface area of the atom will be accessible for water, then $f_\alpha \approx 1$, because $g_w(\sigma;\,\text{protein}) \approx g_w(\sigma;\,\text{bulk})$. On the other hand, f_α at an internal region of protein may be very small or zero, that is, $g_w(\sigma;\,\text{protein}) = 0$, because solvent molecule may not be accessible to such area of protein.

Some reader may be concerned with the water molecules bound inside a cavity of protein, such as an active site (see Chap. 2). Those solvent molecules may give quite large values to $g_w(\sigma;\,\text{protein})$, and thereby to $\zeta_{\alpha\alpha}$. Although the physical origin of the "force" may not be exactly the same as the *friction*, it may be regarded as an *effective friction*, since the velocity of protein atom is considerably *decelerated* due to the solvent molecules.

Some information necessary to calculate $\zeta_{\alpha\alpha}$ may not be obtained thoroughly from experiments, because the atom α is an *imaginary* atom bearing a partial charge. $g_w(\sigma;\text{bulk})$ and $\zeta_{\alpha\alpha,\text{bulk}}$ are examples of such quantities. Nonetheless, the information can be readily obtained from the molecular-dynamics (MD) simulation and/or the RISM/3D-RISM theory (Koneshan et al. 1998; Chong and Hirata 1998, 1999).

Computational Problems in the GLMA: There may be several problems to be concerned with in the actual implementation of the theory to structural fluctuation of a biomolecule in water.

First of those is the time dependence of the force constant matrix $\mathbf{K}$ in Eq. (5.11). Someone may consider that the constant will be changed along the course of the time evolution of the fluctuation. However, it is a baseless concern, because the force constant matrix is invariant as long as the structural fluctuation is progressing within the harmonic free energy surface defined by Eq. (5.6). As been rigorously proved in the preceding chapter, the free energy surface of a biomolecule in water is strictly harmonic as far as the biomolecule is fluctuating due to the thermal motion around the equilibrium structure.

Next question concerns the time dependence of the friction coefficient matrix ζ. The matrix is surely time dependent, because the quantity depends on the structure itself as is evident from its definition, Eqs. (5.23) and (5.24), and the structure changes along the fluctuation. So, if one is trying to pursue the time evolution or rate of a large conformational fluctuation of a biomolecule in water such as *protein folding*, it may require proper care for the time dependence of ζ.

The third problem is concerned with the equilibrium structure $\langle \mathbf{R}_\alpha \rangle$ in Eq. (5.1). The structure is defined by the conformation that makes the free energy defined by Eq. (5.6) minimum. Similar problem has been faced earlier by the community of biophysics in the context of the normal mode analysis (NMA) of a *protein* in *vacuum* (Noguchi and Go 1982). In that case, the authors picked up structural data of a protein from the protein data bank (PDB), and minimized the *potential energy* of the protein *in vacuum* by the technique referred to as "energy minimization" such as the *steepest descent* or/and *conjugate gradient*. We may be able to apply the similar technique to a biomolecule in water, in which the *free energy* defined by Eq. (5.6) is minimized.

In the following section, proposed is a new method to solve the last two problems at the same time.

5.2 Multiscale Dynamics Simulation of a Biomolecule in Water

In the preceding section, an equation to decouple the structural fluctuation of biomolecules in water into vibrational *modes* having different frequencies in a hierarchical order.

In the present section, a theory to realize a multiscale dynamics of protein is proposed based on the generalized Langevin equation combined with the 3D-RISM theory. The idea of normal mode analysis (NMA) is adopted to decouple the dynamics into different modes having respective time scales. In order to decouple the modes, the variance–covariance matrix concerning the structural fluctuation of protein in solution is diagonalized, instead of the Hessian matrix of a harmonic oscillator in vacuum (Hirata and Kim 2016).

Multi-scale Dynamic Algorithm for Solving the GLT/3D-RISM Equation: We consider the dynamics of protein that is taking place around a global minimum of the free energy surface in *solution.* The protein restores its equilibrium structure eventually after any fluctuation around the minimum, small or large, unless the equilibrium conditions such as temperature and pressure are changed. In such a case, we have already derived a generalized Langevin equation for the structural dynamics of protein, which is combined with the 3D-RISM/RISM equation (Eq. 4.52). In the overdamped regime, the equation is written as,

$$\frac{\mathrm{d}^2\Delta\mathbf{R}_\alpha(t)}{\mathrm{d}t^2} = -\sum_\beta K_{\alpha\beta}\cdot\Delta\mathbf{R}_\beta(t) - \sum_\beta \zeta_{\alpha\beta}\Delta\mathbf{V}_\beta + \mathbf{W}_\alpha(t) \tag{5.25}$$

In Eq. (5.25), the Greek suffices specify atoms in protein. $\Delta\mathbf{R}_\alpha(t)$ is the deviation of atom α in protein from its equilibrium position, defined by, $\Delta\mathbf{R}_\alpha(t) = \mathbf{R}_\alpha(t) - \langle\mathbf{R}_\alpha\rangle$, where $\langle\mathbf{R}_\alpha\rangle$ means the statistical average at equilibrium, and $\mathbf{P}_\alpha(t) = M_\alpha \mathrm{d}\mathbf{R}_\alpha/\mathrm{d}t$. $\zeta_{\alpha\beta}$ is the friction coefficient of atom α, and $\mathbf{W}_\alpha(t)$ is the random force. The matrix $\mathbf{L}$ is the variance–covariance matrix concerning the structural fluctuation of atoms, defined by,

$$\mathbf{L} \equiv \langle\Delta\mathbf{R}\Delta\mathbf{R}\rangle \tag{5.26}$$

The essential feature of Eq. (5.25) lies in the first term in the right-hand-side, that takes a form of the Hookian restoring force with $k_\mathrm{B}T\mathbf{L}^{-1}$ as the *force constant* or the *Hessian.* Namely, Eq. (5.25) takes the form of a Langevin harmonic oscillator. It is worthwhile to remember that the harmonic nature of the equation is not originated from the truncation of Taylor expansion of the potential energy at the second order of the displacement as is in the case of NMA, but from the projection of all other variables in the phase space onto the dynamic variables including the atomic displacement of protein (see Chap. 4). The projection changes the dynamics from deterministic to stochastic or diffusive ones, that is governed by the fluctuation–dissipation theorem. It is this feature that we can take advantage in order to decouple the modes of protein dynamics.

We decouple the different modes of dynamics of protein in solution based on the GLMA presented in Sect. 5.1. Firstly, the *Hessian* or *force constant* in Eq. (5.25) is identified as the second derivative of the free energy surface including the solvation free energy, not of the mechanical potential energy surface, that is,

$$k_\mathrm{B}T\left(\langle\Delta\mathbf{R}\Delta\mathbf{R}\rangle^{-1}\right)_{\alpha\beta} = \frac{\partial^2 F(\{\mathbf{R}\})}{\partial\mathbf{R}_\alpha\partial\mathbf{R}_\beta} \tag{5.27}$$

where $F(\{\mathbf{R}\})$ is the free energy surface of the protein in solution, consisting of the interactions among atoms in protein, $U(\{\mathbf{R}\})$, and the solvation free energy $\mu(\{\mathbf{R}\})$ ($\{\mathbf{R}\}$ represents a set of coordinates of atoms in protein). Secondly, all the eigenvalues of the variance–covariance matrix should not necessarily be positive. Since our dynamics takes place in the global minimum of the free energy surface, the eigenvalues associated with the global structural change should be positive. However, those associated with the local structural change such as the exchange of a hydrogen-bond between an amino-acid residue and a water molecule, can be negative, which correspond to a transient state among local free energy minima. We expect that only few largest eigenvalues associated with the collective modes should be positive in order for the protein to stay in the global minimum. In that respect, our approach is not just NMA in its narrow meaning, but rather a general principal-axis (or component) analysis on the quadratic free energy surface.

In the actual implementation of the theory to the GLT/3D-RISM simulation, an optimum size of time step should be carefully chosen to meet the following two requirements: firstly, the time step should be as large as possible in order to maximize the sampling of the free energy surface, and secondly, the first few eigenvalues should be positive during the simulation.

We follow the Verlet-type algorithm for the numerical integration of the GLT/3D-RISM equation. The Verlet-type algorithm reads,

$$\begin{aligned}\Delta\mathbf{R}_\alpha(t+\Delta t) &= 2\Delta\mathbf{R}_\alpha(t) - \Delta\mathbf{R}_\alpha(t-\Delta t) + \frac{\partial^2\Delta\mathbf{R}_\alpha(t)}{\partial t^2}\Delta t^2 \\ \Delta\mathbf{v}_\alpha &\equiv \frac{\Delta\mathbf{R}_\alpha(t+\Delta t) - \Delta\mathbf{R}_\alpha(t-\Delta t)}{2\Delta t}\end{aligned} \tag{5.28}$$

From Eqs. (5.25) and (5.28), one gets

$$\begin{aligned}\Delta\mathbf{R}_\alpha(t+\Delta t) = {} & 2\Delta\mathbf{R}_\alpha(t) - \Delta\mathbf{R}_\alpha(t-\Delta t) \\ & - \left[\sum_\beta \left(K_{\alpha\beta}\Delta\mathbf{R}_\beta + \zeta_{\alpha\beta}\Delta\mathbf{V}_\beta\right)\right]\Delta t^2 \\ & + (\text{Random Force})\end{aligned} \tag{5.29}$$

The transformation of the displacement vectors in the Cartesian space to those in the principal component space is expressed by,

$$\Delta\mathbf{R}_\alpha(t) = \sum_{m=1}^{3N-6} \mathbf{u}_{\alpha m} Q_m(t) \tag{5.30}$$

where $u_{\alpha m}$ is the eigenvector which diagonalizes the *Hessian* matrix $k_B T\mathbf{L}^{-1}$. Q_i is the principal component variable which plays the role of "amplitude" of a "fluctuation" in the principal component space. Substituting Eq. (5.30) into Eq. (5.29) gives rise to

$$\sum_{\alpha=1}^{3N-6} \mathbf{u}_{\alpha m} Q_m(t+\Delta t) = 2\sum_{\alpha=1}^{3N-6} \mathbf{u}_{\alpha m} Q_m(t) - \sum_{m=1}^{3N-6} \mathbf{u}_{\alpha m} Q_m(t-\Delta t)$$
$$-\left[\sum_{\beta}\left(K_{\alpha\beta}\sum_{m=1}^{3N-6} \mathbf{u}_{\alpha m} Q_m(t) + \zeta_{\alpha\beta}\sum_{m=1}^{3N-6} \mathbf{u}_{\alpha m} \dot{Q}_m(t)\right)\right]\Delta t^2 \tag{5.31}$$

In the above expression, we have ignored the random force term, because it should be taken care of in the process of temperature control anyway in actual MD simulations in real space.

From Eq. (5.31), one finds the equation of motion in the Generalized-Langevin-Mode (GLM) space as,

$$Q_m(t+\Delta t) = 2Q_m(t) - Q_m(t-\Delta t) - \Lambda_m \dot{Q}_m(t) - \Gamma_m \dot{Q}_m(t), \tag{5.32}$$

where Λ_m and Γ_m are the eigen values of the Hessian and friction matrices defined by Eq. (5.12), respectively.

The equation allows us to scale the displacements in the principal component space during a time step Δt depending on the mode. A collective mode such as the hinge bending motion has large Q_m, while a small vibration such as the bond-stretching motion has small Q_m. So, an appropriate Δt should be chosen according to the mode, the displacements in the principal component space are quite different between the two. For example, typical frequency of the hinge bending motion, which is essentially diffusive, should be almost zero or at least no greater than 10 cm^{-1}, while that of bond-stretching is few thousand wavenumbers. So, the difference is few hundred times in this example. It should be noted that the motions are completely orthogonal in the principal component space, provided that the *Hessian* matrix is properly diagonalized. Therefore, we can sum up the modes, based on Eq. (5.30), to reconstruct a displacement vector in the Cartesian-coordinate space after time Δt without destroying the real space dynamics, such as *overlap of two atoms* in protein.

The protocol to realize the dynamics of protein based on Eq. (5.31) is as follows.

1. Prepare the displacement vectors at the two times, t and $t - \Delta t$, separated by Δt.
2. Calculate the Hessian matrix $K_{\alpha\beta}$ (Eq. 5.27) based on the 3D-RISM/RISM theory.
3. Calculate the friction matrix $\zeta_{\alpha\beta}$ according to the recipe explained in the preceding Sect. 5.1.
4. Diagonalize $\mathbf{K}$ to get the principal component variables Q_m as well as its time derivative and the eigenvectors $\mathbf{u}_{\alpha m}$. Diagonalize also the friction coefficient matrix in order to find that correspond to principal component variables.
5. Scale the principal component variables Q_m and $\mathrm{d}Q_m/\mathrm{d}t$ by the time step Δt^2.
6. Reconstruct the displacement vector $\Delta\mathbf{R}_\alpha$ in the Cartesian-coordinate space as a linear combination of the scaled principal component variables.

7. Perform the ordinary MD simulation in atomic level in Cartesian coordinate space to find $\Delta \mathbf{R}_\alpha(t + \Delta t)$ at time $t + \Delta t$.

References

Chong S-H, Hirata F (1998) Dynamics of solvated ion in polar liquids: an interaction-site-model description. J Chem Phys 108(17):7339–7349
Chong S-H, Hirata F (1999) Dynamics of ions in liquid water: an interaction-site-model description. J Chem Phys 111(8):3654–3667
Hirata F, Kim B (2016) Multi-scale dynamics simulation of protein based on the generalized Langevin equation combined with 3D-RISM theory. J Mol Liq 217:23
Kim B, Hirata F (2013) Structural fluctuation of protein in water around its native state: a new statistical mechanics formulation. J Chem Phys 138:054108–054111
Koneshan S, Rasaiah JC, Lynden-Bell RM (1998) Sovent structure, dynamics, and ion mobility in aqueous solutions at 25 C. J Phys Chem B 102:4193–4204
Kovalenko A, Hirata F (1998) Three-dimensional density profiles of water in contact with a solute of arbitrary shape: a RISM approach. Chem Phys Lett 290:237–244
Kovalenko A, Hirata F (1999) Self-consistent description of a metal-water interface by the Kohn–Sham density functional theory and the three-dimensional reference interaction site model. J Chem Phys 110:10095–10112
Lamm G, Szabo A (1986) Langevin mode of macromolecules. J Chem Phys 85:7334
Miyata T, Hirata F (2007) Combination of molecular dynamics method and 3D-RISM theory for conformational sampling of large flexible molecules in solution. J Comput Chem 29:871–882
Noguchi T, Go N (1982) Nature 296:776
Omelyan I, Kovalenko A (2015) MTS-MD of biomolecules steered with 3D-RISM-KH mean solvation forces accelerated with generalized solvation force extrapolation. J Chem Theory Comput 11(4):1875–1895
Wang MC, Uhlenbeck GE (1945) On the theory of Brownian motion II. Rev Mod Phys 17:323–342
Yoshida N, Hirata F (2006) A new method to determine electrostatic potential around macromolecule in solution from molecular wave functions. J Comput Chem 27(4):453–462
Yu H-A, Karplus M (1988) A thermodynamics analysis of solvation. J Chem Phys 89:2366

Chapter 6
Spectral Analysis of a Biomolecule in Water by GLMA

In the previous chapter, we have developed a statistical mechanics theory referred to as the generalized Langevin mode analysis (GLMA) that enables us to analyze the structural fluctuation of a biomolecule in aqueous solutions. The present chapter is devoted to a tentative application of the theory to explain an experimental result concerning the structural fluctuation.

In Sect. 6.1, we just make a brief remark on the motivation to carry out the GLMA analysis on the low-frequency mode of a biomolecule in water. Section 6.2 is devoted to review the theoretical analysis of the low-frequency spectrum of an alanine dipeptide in water, which is carried out based on GLMA and the molecular dynamics (MD) simulation. The MD simulation was performed on the free energy surface composed of the interatomic potential energy of protein and the solvation free energy. The theoretical results are compared with the spectrum reported by Wynne and his coworkers based on the Rahman-Induced Kerr Effect Spectroscopy (RIKES) (Giraud and Wynne 2002).

6.1 Introductory Remark

As may be readily imagined, the structural fluctuation of a biomolecule has a very wide dynamic range from nearly zero to more than a thousand wavenumbers. The wavenumbers related to the bond lengths and bond angles are surely quite large. On the other hand, wavenumbers related to the collective modes such as the hinge bending are small (McHaourab et al. 1997). It is readily imagined that the collective modes related to the protein folding/unfolding would have quite small wavenumber, since the range of fluctuation extends over the entire molecule.

Such fluctuations of protein structure have been observed by means of a variety of spectroscopy such as infrared, ultraviolet, circular dichroism, Rahman scattering, and so on. However, the dynamic range of fluctuation, that can be measured naturally

F. Hirata, *Molecular Theory of Fluctuation in Life Phenomena*,
https://doi.org/10.1007/978-981-96-3364-7_6

by those conventional methods, was rather limited to that of higher wavenumber, greater than ~ a few hundred cm^{-1}.

The most important restriction for those conventional methods is an anisotropy of electric as well as magnetic polarization of a biomolecule in water. As been clarified theoretically in the preceding chapter, the structural fluctuation of a biomolecule, represented by the variance–covariance matrix $\langle \Delta \mathbf{R} \Delta \mathbf{R} \rangle$, is extremely *anisotropic*. Therefore, it is the least requirement for any method of spectroscopy to be able to account for such *anisotropy* of polarization in one way or the other.

A spectroscopic technique that probes the "electro-optical Kerr effect" is one of those methods which reflect such anisotropy of the electric polarization due to the inhomogeneous fluctuation of a biomolecule in water. The effect was first discovered by Kerr in 1875, but it is rather recent that the spectroscopic method to detect the effect in sufficient accuracy has been developed (Kerr 1875).

Recently, Wynne and his coworkers have developed a method referred to as Rahman-Induced Kerr Effect Spectroscopy (RIKES) based on the optical heterodyne method. Based on the method, the group has been providing experimental data concerning the spectral data with respect to the structural fluctuation of a biomolecule in water for a quite wide dynamic range including the collective modes (Giraud and Wynne 2002).

6.2 Generalized Langevin Mode Analysis (GLMA) of Alanine Dipeptide in Water

The observed spectrum is a quantity averaged over the molecules, the conformation of which is fluctuating in time as well as in space. Therefore, we have carried out a *molecular dynamics (MD) simulation* of an alanine dipeptide in water, before having made the harmonic analysis to calculate the spectrum. It should be noted, however, that the MD simulation carried out here is not a usual one based on the *Newtonian dynamics* on the *energy* surface. It is because such dynamics will not produce a trajectory which gives the quadratic surface for the harmonic analysis, as has been clarified in the preceding chapter.

The simulation we have carried out is dynamics on the *free energy* surface described by Eq. (4.59), or

$$F(\{\mathbf{R}\}) \equiv U(\{\mathbf{R}\}) + \Delta\mu(\{\mathbf{R}\}),$$

where $U(\{\mathbf{R}\})$ is the interaction energy among atoms in protein, and $\Delta\mu(\{\mathbf{R}\})$ is the solvation free energy. The free energy surely produces a trajectory that meets the condition for a harmonic analysis (Omelyan and Kovalenko 2015).

The interaction energy among atoms in a biomolecule was calculated based on the standard molecular mechanics potential-energy surface expressed by

$$\begin{aligned}
U(\{\mathbf{R}\}) &= E^{\text{bond}} + E^{\text{ang}} + E^{\text{tors}} + E^{\text{LJ}} + E^{\text{elec}} \\
&= \sum_{\text{bond},i} K_i^r \left(r_i - r_i^0\right)^2 + \sum_{\text{angle},i} K_i^\theta \left(\theta_i - \theta_i^0\right)^2 \\
&\quad + \sum_{\text{tors},i} V_i^\varphi \left\{1 + \cos\left(n_i \varphi_i - \varphi_i^0\right)\right\} \\
&\quad + \sum_{\substack{\text{non-bonded}\\ \text{pair},ij}} \left[\varepsilon_{ij}\left\{\left(\frac{\sigma_{ij}}{r_{ij}}\right)^{12} - \left(\frac{\sigma_{ij}}{r_{ij}}\right)^{6}\right\} + \frac{z_i z_j}{r_{ij}}\right].
\end{aligned} \tag{6.1}$$

The solvation free energy $\Delta\mu(\{\mathbf{R}\})$ and its first and second derivatives are calculated based on Eqs. (2.66), (5.16), and (5.17) appeared in previous chapters.

$$\Delta\mu = \rho k_B T \sum_a \int \left[\frac{1}{2} h_a(\mathbf{r})^2 - c_a(\mathbf{r}) - \frac{1}{2} h_a(\mathbf{r}) c_a(\mathbf{r})\right] d\mathbf{r}$$

$$\frac{\partial}{\partial \mathbf{R}_\alpha} \Delta\mu(\{\mathbf{R}\}) = \sum_i \rho_i \int \frac{\partial u_i^{uv}(\mathbf{r}; \{\mathbf{R}\})}{\partial \mathbf{R}_\alpha} g_i^{uv}(\mathbf{r}; \{\mathbf{R}\}) d\mathbf{r}$$

$$\begin{aligned}
\frac{\partial^2}{\partial \mathbf{R}_\alpha \partial \mathbf{R}_\beta} \Delta\mu(\{\mathbf{R}\}) &= \sum_i \rho_i \int \frac{\partial^2 u_i^{uv}(\mathbf{r}; \{\mathbf{R}\})}{\partial \mathbf{R}_\alpha \partial \mathbf{R}_\beta} g_i^{uv}(\mathbf{r}; \{\mathbf{R}\}) d\mathbf{r} \\
&\quad + \sum_i \rho_i \int \frac{\partial u_i^{uv}(\mathbf{r}; \{\mathbf{R}\})}{\partial \mathbf{R}_\alpha} \frac{\partial}{\partial \mathbf{R}_\beta} g_i^{uv}(\mathbf{r}; \{\mathbf{R}\}) d\mathbf{r}.
\end{aligned}$$

Molecular Dynamics Simulation of an Alanine Dipeptide on the Free Energy Surface: Skipping all the technical details of the simulation, the author just shows the trajectory of alanine dipeptide projected on the two-dimensional surface spanned by the two dihedral angles (ψ_1, φ_2) shown in Fig. 6.1 (Sugita and Hirata 2023).

The major peak of the distribution is seen around $\psi_1 = 150°$ and $\varphi_2 = -60°$, marked by (ii) in Fig. 6.1b, which corresponds roughly to the *trans-gauche* isomer. A minor peak of the distribution is seen around $\psi_1 = 150°$ and $\varphi_2 = -150°$, which is assigned roughly to the *trans-trans* isomer.

An interesting feature is observed in the trajectory, that is worthwhile to be mentioned. That is the few points seen between the two major conformations mentioned above. The result suggests that those points are manifestation of the transient between the two major isomers stated above. Those transient structures will contribute to the spectrum in the negative wavenumber region, because the curvature of the free energy surface in the corresponding regions is negative.

However, it is not whole story. The points between the two local minima in the dihedral-angle space represent local "modes" or "higher-frequency mode" in terms of the harmonic analysis. Those modes of oscillation are *localized* around one of the dihedral angles. On the other hand, there are oscillational modes that extend over the entire molecule. Those modes are referred to as "collective modes". The

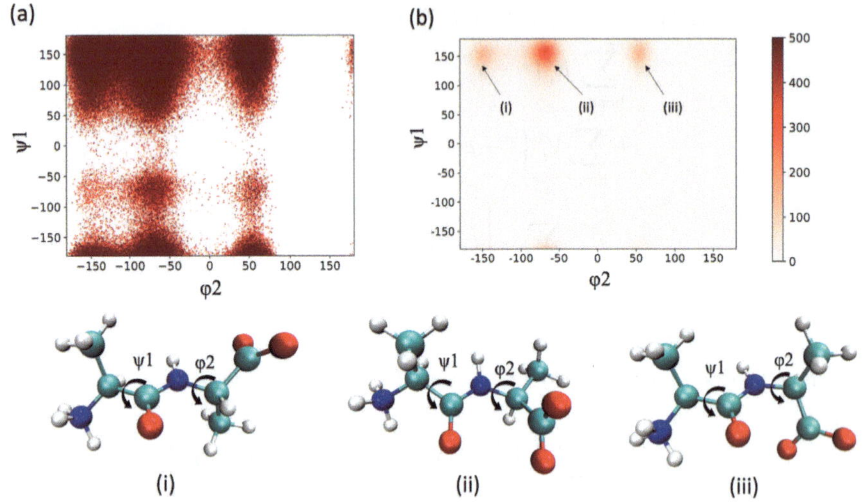

Fig. 6.1 Trajectory projected onto the dihedral-angle (ψ_1, φ_2) space: **a** the snapshots at every 80 steps, **b** the distribution of the trajectory. The molecular pictures depicted under the two figures **a** and **b** illustrate the conformations corresponding to the snapshots (i), (ii), and (iii) in **b**

collective mode with the least wavenumber will never produce a spectrum in the negative frequency region, as far as the system is in an equilibrium state. Why is it so? The collective mode of a biomolecule corresponds to a global structure, such as native conformation of protein. The free energy surface corresponding to such a structure does not have a negative curvature, unless otherwise the structure will change entirely, such as the native to denatured conformational-change.

Therefore, the points between the two minima in the dihedral-angle space are manifestations of two physical processes. One of those is a transient state between the two local minima in the dihedral-angle space. It is the mode which causes the negative frequency in the spectrum, because it indicates that the molecule is crossing over the free energy barrier between the two isomers corresponding to the different dihedral angles. The other is a conformation in the collective mode with the lowest wavenumber. Such a mode will never contribute to the spectrum with a negative frequency, because curvature of the free-energy well corresponding to such a mode will be positive all the way in the trajectory.

Spectrum from Multiple Snapshots: The wavenumber spectrum of an alanine dipeptide in water is shown in Fig. 6.2. The spectrum was obtained by diagonalizing the Hessian matrix, $K_{\alpha\beta}$, and averaging over 1000 snapshots. The snapshots were sampled from an 80 ns 3D-RISM/MD trajectory, which are evenly spaced. The structure of each snapshot was not the one minimized on the free energy surface.

The spectrum depicted in Fig. 6.2 can be discussed in the four different regions: (i) the region greater than ~ 2700 cm^{-1}, (ii) the region between ~ 500 and ~ 1800 cm^{-1}, (iii) the region less than ~ 500 cm^{-1}, and (iv) the region of negative wavenumber.

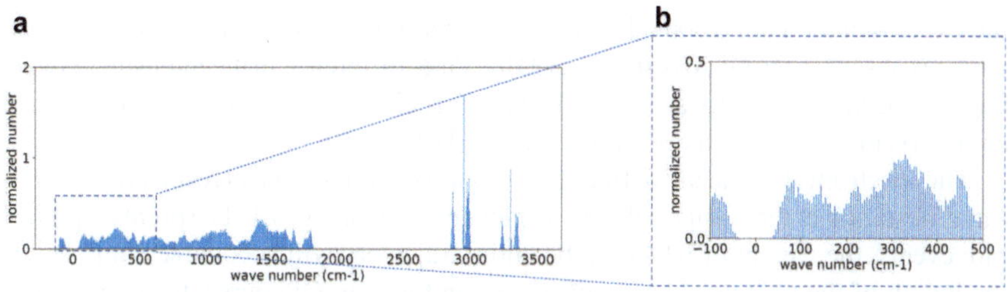

Fig. 6.2 Wavenumber spectrum of an alanine dipeptide in water: **a** the spectrum calculated from the 1000 snapshots, evenly spaced, without minimizing the free energy; **b** the spectrum calculated from the 100 snapshots, evenly spaced, after minimizing the free energy. The inset is exploded view of the spectrum in the range of wavenumber from − 100 to 500 cm^{-1}

The regions greater than ~ 2700 cm^{-1} are attributed apparently to stretching and bending of chemical bonds related to hydrogen atoms. The spectrum between ~ 500 and ~ 1800 cm^{-1} is apparently assigned to the local oscillations, stretching and bending, in which the heavy atoms such as C, O, and N are involved. Those vibrational modes related to heavy atoms are essentially determined by the model of force field adopted by the simulation program. Those are not much of interest for the structural fluctuation of biomolecules, especially for the collective modes of structural fluctuation. It is the regions (iii) and (iv) that are closely related to the structural fluctuation. We focus our attention on that wavenumber region.

The spectrum less than 500 cm^{-1} is shown by histogram in Fig. 6.3. Corresponding experimental (RIKES) data by Wynne and his coworkers are also depicted (Sugita and Hirata 2023; Giraud and Wynne 2002).

Although there are apparent differences observed between the two results, there is a common feature in the two spectra. The two spectra have four peaks between the wavenumbers, 0–500 cm^{-1}, which are relatively close each other, that is, ~ 90, 250, 370, and 450 cm^{-1} in the RIKES spectra, while ~ 90, ~ 240, ~ 319, and ~ 450 cm^{-1} in the GLMA spectra.

There are marked differences between the two spectra in the following respects: (1) the negative frequency region observed in the GLMA result, which is absent in

Fig. 6.3 Comparing the low-frequency spectrum from GLMA with that from RIKES

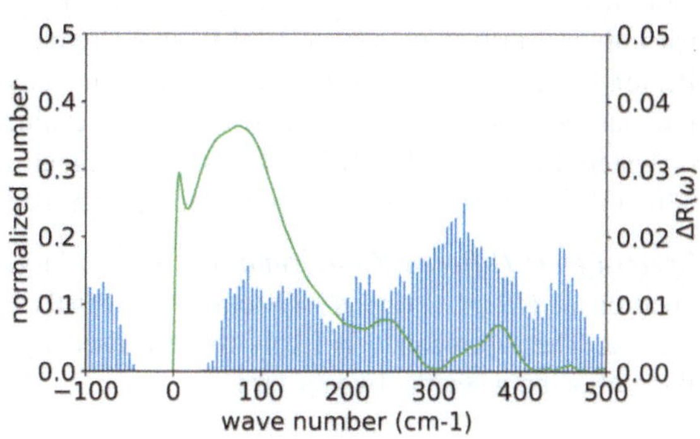

the experimental data; (2) small subpeak around 0 cm^{-1} seen in the RIKES result, which is absent in the GLMA spectra; (3) the large difference in the intensity between the two spectra, especially around 0–100 cm^{-1} region; (4) relatively large difference, about 50 cm^{-1}, in the peak positions around 319 cm^{-1}.

The spectrum at the negative frequency region in the theoretical result is apparently a manifestation of the points in the $\phi - \psi$ plot depicted in Fig. 6.1. The points represent rare events, the weight of which in the entire trajectory will become less and less as the length of the trajectory becomes longer and longer. Of course, the contribution from such rare events disappears in the real world of experiments.

From (2) to (4) differences listed above, between the experiment and the theory, are caused essentially by the difference in the physics related to the two analyses. According to Giraud and Wynne, the experimental spectrum was acquired by subtracting the spectrum of pure solvent from the entire spectrum of solution, including both solute and solvent molecules (Giraud and Wynne 2002). Therefore, the intensity shown in the figure includes two contributions, one from the solute and the other from water molecules that are interacting with the solute. The analysis suggests that the contribution from the water molecules interacting with the solute is not removed from the entire spectrum. It also suggests that the small subpeak observed around 0 cm^{-1} is likely to be assigned to the *diffusive motion* of water molecules interacting with the dipeptide. On the other hand, any spectrum related to water molecules is not included in the theoretical analysis, although the contribution from solute–solvent interaction is included implicitly in the spectrum of the solute. It may be the reason why the theoretical result does not show spectrum around zero wavenumber. The experimental spectra may include also the spectrum from the translational and rotational diffusion of the dipeptide, while the contribution from such degrees of freedom is entirely removed from the theoretical analysis.

The large difference in the intensity between the two spectra, especially those around ~ 100 cm^{-1} region, is attributed to the same reason. The RIKES spectrum in the region includes contributions from water molecules around the solute. Those water molecules are likely to be involved in interaction with the solute molecule, especially, via hydrogen bond. They are likely to be involved in the rotational and translational oscillation in lower frequency modes. So, it is suggested that the large intensity covering the 100 cm^{-1} region is assigned to the intermolecular oscillatory motion of water molecules interacting with the solute. There is a simulation study that partially supports the assignment of the low-frequency spectrum, suggested above. Person et al. have carried out a MD simulation study for water molecules around a monoatomic solute, Na^+, K^+, Ne, A, and Xe, to evaluate the low-frequency spectrum of water (Person et al. 2017). They have assigned the large intensity around 0 ~ 300 cm^{-1} to the rot-vibrational mode of water molecules.

Spectra from Different Conformations: Rather large difference is seen in the peak positions, ~ 319 cm^{-1} versus ~ 370 cm^{-1} between the theoretical and the experimental spectra. The discrepancy is likely to be caused by the difference in the mode of fluctuation between the two spectra.

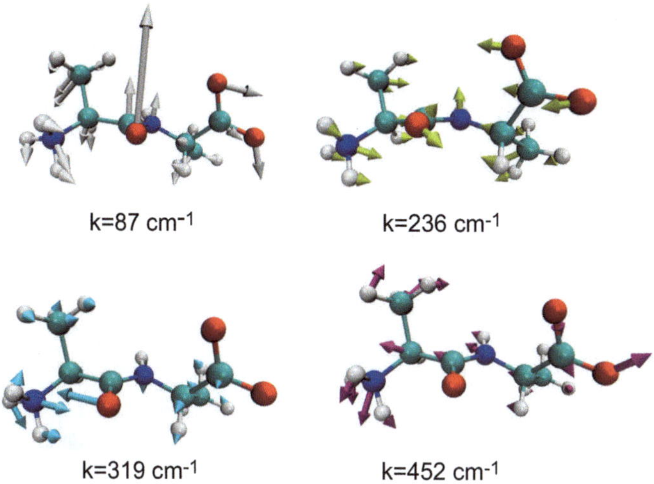

Fig. 6.4 Oscillatory modes corresponding to the peaks of wavenumber spectrum

The conformation and the oscillatory mode of the dipeptide corresponding to the peak positions in the GLMA spectrum are illustrated in Fig. 6.4. The direction and amplitude of the oscillation of each atom are depicted by thick arrows.

As can be observed, all the modes have a collective character, more or less. It indicates that the oscillations are not limited to a bond or an angle, but spread over the entire molecule. For examples, the lowest frequency mode, or 87 cm^{-1}, looks like a "hinge-bending" motion around the C=O carbonyl bond, because the central carbonyl group and the two terminal groups are oscillating in the opposite phases. On the other hand, the mode assigned to 452 cm^{-1} looks more localized around the N-terminus group. It may explain why the frequency is relatively higher.

The mode assigned to 319 cm^{-1} is worthwhile to draw a special attention. In the mode, the carbonyl oxygen (red sphere) and the amide nitrogen (blue sphere) are oscillating in the opposite phase. The picture is suggesting that a water molecule is making a bridge between the two atoms through a hydrogen bond. The result suggests a study to analyze the solvation structure of the molecule in the oscillatory mode assigned to $k = 319$ cm^{-1} in order to make the physics clear.

Depicted in Fig. 6.5 is the radial distribution functions (RDF) of water molecules around the carbonyl oxygen and the amide nitrogen of the dipeptide. As can be observed, there is a sharp peak at $r \sim 1.8$ A in the O(peptide)–H(water) RDF in Fig. 6.5c. It clearly indicates that the carbonyl oxygen and the water molecule are hydrogen-bonded through the hydrogen atom. On the other hand, the sharp peak at $r \sim 2.8$ A in the N(peptide)–O(water) RDF in Fig. 6.5a is a manifestation of a strong electrostatic interaction between the amide nitrogen and the water oxygen. The two sharp peaks observed in RDFs are clear evidences to prove that the carbonyl oxygen and the amide nitrogen are bridged by a water molecule. Shown at right in Fig. 6.5d

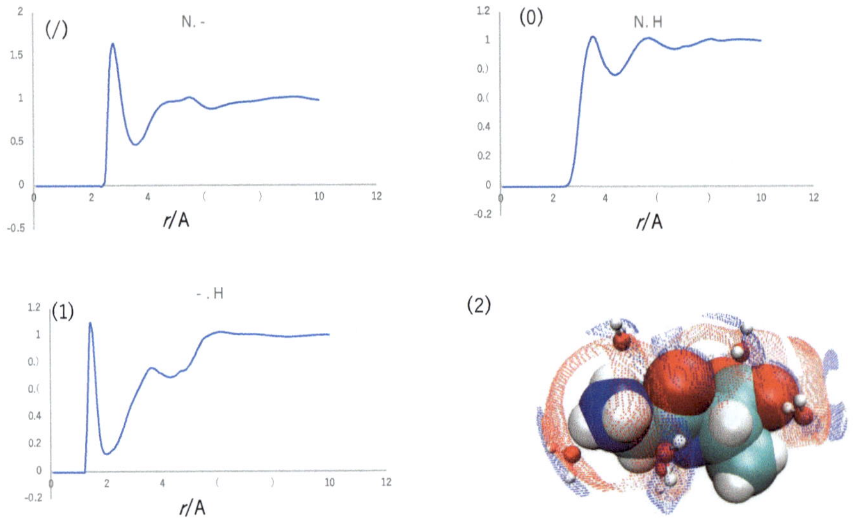

Fig. 6.5 Radial distribution functions (RDF) of water molecules around the dipeptide in the fluctuational mode with 319 cm^{-1}. The figure depicted by the red and blue spots in **d** is the 3D distribution of the H and O atoms of water molecules around the dipeptide: red, oxygen; blue, hydrogen. Note that the distribution of the hydrogen atom extends near the carbonyl oxygen of the peptide, which creates the hydrogen-bond peak at $r \sim 1.5$ A in the RDF of O(peptide)–H(water) pair in **c**

is an image that illustrates the *water bridge*. It is suggested that the water bridge is the origin of the oscillatory mode assigned to 319 cm^{-1}.

Now, we are at the position to answer the question raised at the end of the last subsection. Why there is difference in the peak positions in the theoretical and experimental spectra, ~ 319 cm^{-1} versus ~ 370 cm^{-1}, and what is the difference caused by?

A quick answer to the question is that the difference is originated from the physics explored by the two methods. The experimental spectrum includes both the contributions from the dipeptide and water molecules. On the other hand, the theoretical spectrum originates only from the oscillatory modes of the dipeptide. Of course, the theoretical spectrum includes contributions from solvent *implicitly*. The peak in the theoretical spectrum at ~ 319 cm^{-1} is originated mainly by the H(amide)–O(carbonyl) interaction in the peptide, bridged by a water molecule. On the other hand, the experimental spectrum at ~ 370 is a composite band consisting of the contributions from the H(amide)–O(carbonyl) interaction in the peptide and the water molecule bridging the two atoms in the peptide. The frequency of the mode of the water molecule may be higher than that of H(amide)–O(carbonyl) interaction, because the water molecule is bound with the two atoms in the peptide through the two strong interactions.

References

Giraud G, Wynne K (2002) Time-resolved optical Kerr-effect spectroscopy of low-frequency dynamics in di-L-alanine, poly-L-alanine, and lysozyme in solution. J Am Chem Soc 124(41):12110–12111

Kerr J (1875) A new relation between electricity and light: dielectrified media birefringent. Philos Mag 50(332):337–348, 446–458

McHaourab HS, Oh KJ, Fang CJ, Hubbel WL (1997) Conformation of T4 lysozyme in solution. Hinge-bending motion and the substrate-induced conformational transition studied by site-directed spin labeling. Biochemistry 36:307–316

Omelyan I, Kovalenko A (2015) MTS-MD of biomolecules steered with 3D-RISM-KH mean solvation forces accelerated with generalized solvation force extrapolation. J Chem Theory Comput 11(4):1875–1895

Person RAX, Pattni V, Singh A, Kast SM, Heyden M (2017) Signatures of solvation thermodynamics in spectra of intermolecular vibrations. J Chem Theor Comput 13:4467–4481

Sugita M, Hirata F (2023) Realization of the structural fluctuation of biomolecules in solution: generalized Langevin mode analysis. J Comput Chem 44:1031–1039

Chapter 7
Transitions Among Equilibrium States Driven by Fluctuations (Perspective)

In the preceding chapters, we have explored theoretically a variety of fluctuations taking place in life phenomena: density fluctuations of solvent and structural fluctuations of a biomolecule, temporal as well as spatial. The main conclusion of the chapters is that both the structural and density fluctuations are harmonic, and accordingly the free energy surfaces of fluctuations are quadratic.

In the present chapter, we extend the discussion to transitions among equilibrium states, that are driven by structural as well as density fluctuations in biomolecular systems. The transition due to the structural fluctuation is nothing but a *chemical reaction* categorized as *isomerization reactions*. The *protein folding* is the most outstanding example of such an isomerization reaction. Description of chemical reactions in this chapter may be classified into (i) chemical equilibrium (Sect. 7.1), (ii) reaction induced by a chemical/mechanical perturbation (Sect. 7.2), and (iii) rate of reaction (Sect. 7.3), which are intertwined each other. The chemical equilibrium is formulated by the response theory, originally developed by Kubo and his coworkers (Kubo and Toda 1991). Conformational changes of biomolecules due to changes in thermodynamic conditions are described by considering change in thermodynamic variables as a perturbation.

The rate of transition or reaction is described in the framework of the Arrhenius theory concerning the activation-barrier-crossing model. The activation barrier is described by a crossing point of two parabolas which describe the free energy surface of reactant and product states. Chemical reactions associated with recombination of the chemical bonds are out-of-scope of the present chapter. The essential element of such chemical reactions is a significant reorganization of the electronic structure of molecules in water. Such a problem by itself is a challenging problem in the theoretical chemistry. It will be a great perspective to combine the two theories to

The original version of the chapter has been revised. A correction to this chapter can be found at https://doi.org/10.1007/978-981-96-3364-7_8

F. Hirata, *Molecular Theory of Fluctuation in Life Phenomena*,
https://doi.org/10.1007/978-981-96-3364-7_7

explore chemical reactions in a biomolecular system, which are associated with the recombination of chemical-bonds.

As an example of the transition induced by *density fluctuation*, we consider the conductivity of ion channels, mentioned in Introduction (Chap. 1) of the present book. The permeation of an ion through a channel is viewed as a series of barrier crossing along the channel axis, and the rate of barrier crossing is formulated based on the Arrhenius theory (Sect. 7.4).

7.1 Structural Transition of Proteins Induced by Thermodynamic Perturbations

Biomolecules in water change their conformation responding to the thermodynamic environment such as temperature, pressure, and concentration of co-solvent molecules. The *protein folding* and the *B-to-Z transition* of DNA are the most highlighted examples of such conformational change induced by change in thermodynamic conditions.

It has been proved in Chap. 4 that the structural fluctuation of a biomolecule in aqueous solution is *harmonic*, and that the free energy surface is quadratic function of the fluctuation around an equilibrium structure, described by the following equation (Eq. 4.60),

$$F(\{\Delta\mathbf{R}\}) = \frac{1}{2}k_BT\sum_{\alpha,\beta}\Delta\mathbf{R}_\alpha\cdot K_{\alpha\beta}\cdot\Delta\mathbf{R}_\beta. \tag{7.1}$$

In the equation, $\Delta\mathbf{R}_\alpha$ represents the structural fluctuation defined by

$$\Delta\mathbf{R}_\alpha = \mathbf{R}_\alpha - \langle\mathbf{R}_\alpha\rangle \tag{7.2}$$

where $\langle\mathbf{R}_\alpha\rangle$ denote the equilibrium structure. Accordingly, the probability distribution of the structural fluctuation becomes Gaussian, or

$$\begin{aligned} w(\{\Delta\mathbf{R}\}) &= C\exp[-F(\{\Delta\mathbf{R}\})/k_BT] \\ &= \sqrt{\frac{\|K\|}{(2\pi)^{3N}}}\exp\left[-\frac{1}{2}\sum_\alpha\sum_\beta\Delta\mathbf{R}_\alpha K_{\alpha\beta}\Delta\mathbf{R}_\beta\right] \end{aligned} \tag{7.3}$$

where $\|K\|$ is the determinant of the matrix $K_{\alpha\beta}$, which is the inverse matrix of the variance–covariance matrix defined by Eq. (4.17), or

$$L_{\alpha\beta} \equiv \langle\Delta\mathbf{R}\Delta\mathbf{R}\rangle_{\alpha\beta} \tag{7.4}$$

In the present chapter, we consider phenomena in which biomolecules change its equilibrium conformation $\langle \mathbf{R}_\alpha \rangle$ responding to a perturbation induced by a change in external conditions such as thermodynamic environments, chemical and/or mechanical perturbation, and so on. Such a change of the equilibrium structure is nothing but a *chemical reaction* referred to as "isomerization reaction" in chemistry.

In the following, we formulate a theory to describe such conformational transition induced by changes in thermodynamic conditions based on the response theory in the statistical mechanics (Hirata and Akasaka 2015; Friedman 1985; Kubo and Toda 1991).

Linear and Nonlinear Response Theories: Let's think of an aqueous solution of protein in ambient condition, or room temperature and one atmosphere. The protein is supposed to be in its native conformation. Assume the thermodynamic condition is changed by applying some perturbation such as *heat* or *pressure*. The structure of protein as well as its free energy should be changed due to the perturbation. The structural change can be predicted based on the variational principle as follows.

Firstly, we add an energy change, proportional to the perturbation ($\mathbf{f}$), to the free energy surface defined in Eq. (7.1).

$$F(\{\Delta \mathbf{R}\}) = \frac{1}{2} k_{\mathrm{B}} T \sum_{\alpha,\beta} \Delta \mathbf{R}_\alpha \cdot K_{\alpha\beta} \cdot \Delta \mathbf{R}_\beta - \sum_{\alpha} \Delta \mathbf{R}_\alpha \cdot \mathbf{f}_\alpha \tag{7.5}$$

where $\mathbf{f}_\alpha$ and $\Delta \mathbf{R}_\alpha$ are the perturbation and the displacement of atoms caused by the perturbation, respectively. Then, apply the variational principle to Eq. (7.5) to find a linear response expression describing structural change induced by the perturbation (The expression has been originally derived by Ikeguchi et. al. in a different way.) (Ikeguchi et al. 2005)

$$\langle \Delta \mathbf{R}_\alpha \rangle_1 = (k_{\mathrm{B}} T)^{-1} \sum_{\beta} \left\langle \Delta \mathbf{R}_\alpha \Delta \mathbf{R}_\beta \right\rangle_0 \cdot \mathbf{f}_\beta \tag{7.6}$$

The linear response theory, Eq. (7.6), can be applied to the structural change induced by a thermodynamic perturbation such as pressure, temperature, and denaturant such as urea.

Such a structural change of protein due to a perturbation is illustrated in Fig. 7.1. Actual structural change of protein may not be drawn by such a simple picture as Fig. 7.1, because a protein usually has ~ 10^4 atoms. Therefore, the probability distribution of structure should be drawn in the conformational space spanned by ~ 10^4 coordinates, which is impossible of course. So, the figure is just an image.

The linear response theory may be applied to a case in which the perturbation does not cause a significant change in the variance–covariance matrix $\left\langle \Delta \mathbf{R}_\alpha \Delta \mathbf{R}_\beta \right\rangle$. The protein folding is the best example in which the linear response theory may not be applied, since the variance–covariance matrix will be dramatically reduced upon folding into the native conformation. However, the concept stated above can be extended to a non-linear regime, in which the response function varies with the

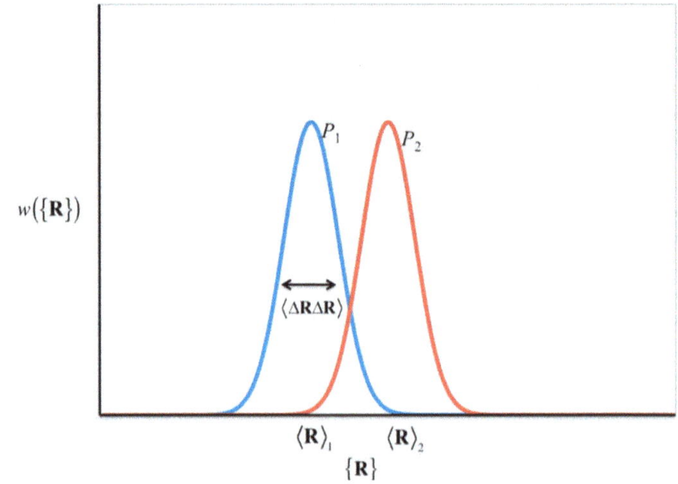

Fig. 7.1 Image of structural change induced of protein by a perturbation, as a shift of Gaussian distribution: $\{\mathbf{R}\}$ denotes a set of Cartesian coordinates of atoms in protein, representing its structure

thermodynamic perturbation. The extension can be made in terms of the *analytical continuation* (Hirata and Akasaka 2015).

We divide the perturbation and the response into several steps in such a way that each step stays within a *convergence radius*, where the linear response theory is valid. The process can be formulated as

$$\langle \Delta\mathbf{R}_\alpha \rangle_{j+1} = \left(k_B T_j\right)^{-1} \sum_\beta \left\langle \Delta\mathbf{R}_\alpha \Delta\mathbf{R}_\beta \right\rangle_j \cdot \mathbf{f}_\beta \tag{7.7}$$

where j starts from zero. In the equation, the variance–covariance matrix at the (j + 1)th step should be calculated based on the protein coordinates in the jth step. (Although temperature has suffix j, or T_j, it should stay constant unless temperature itself is the thermodynamic perturbation.) With a proper care made for "scheduling" of increasing, or decreasing, the perturbation, we may be able to get successive Gaussian distributions for varying thermodynamic conditions with corresponding $\langle \Delta\mathbf{R}_\alpha \rangle$ and $\left\langle \Delta\mathbf{R}_\alpha \Delta\mathbf{R}_\beta \right\rangle$. An *image* of such a process is illustrated in Fig. 7.2. In the figure, $w(\{\mathbf{R}\})$ denotes the probability distribution of the structure of protein, represented by $\{\mathbf{R}\}$.

The figure describes the protein folding (or unfolding) as a process in which the Gaussian distribution (the average structure and the variance covariance matrix) shifts reversibly according to a thermodynamic perturbation described by P_i.

The theory views the protein structure as a *Gaussian* distribution, the peak position and the half width of which correspond, respectively, to the equilibrium structure and the variance–covariance matrix. Accordingly, the protein folding is viewed as the shift of the peak position, and the change of the half width of the Gaussian distribution. Is there any experimental evidence to confirm the theoretical view?

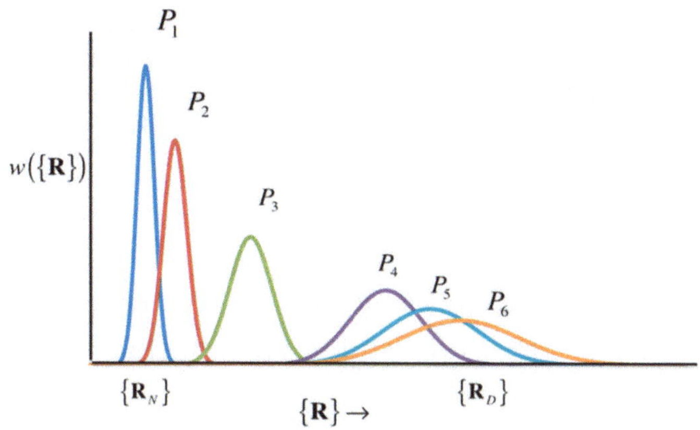

Fig. 7.2 Illustration of the shift of Gaussian distribution of protein structure along the pathway from native to denatured states, or vice versa. It should be noted that the figure is just a schematic presentation. A real distribution is composed of many Gaussian functions with different half width

Yes, there are! Those are the results of the pressure NMR experiment carried out by Akasaka and his coworkers. The authors have developed a unique method to realize the fluctuation of protein in solutions. The idea is based on their finding that the NMR chemical-shift of amino acids changes upon increasing pressure. The change becomes more significant as the pressure is enhanced. Based on the finding, Akasaka has proposed a hypothesis that an equilibrium structure at higher pressure should correspond to one of fluctuated structures at the ambient pressure. It suggests that the fluctuated states of protein can be realized by changing pressure as an *order parameter* (Akasaka 2006).

Description of Thermodynamic Perturbations: We define the force acting on an atom of protein in solution by the derivative of the Helmholtz free energy of the system with respect to the coordinates of the atom,

$$\mathbf{f}_\alpha = -\left(\frac{\partial F_\mathrm{s}}{\partial \mathbf{R}_\alpha}\right) \tag{7.8}$$

where $\mathbf{R}_\alpha$ and F_s denote the αth atom in protein and the Helmholtz free energy of the *entire system*, respectively. Note that the notation F_s is used to distinguish it from free energy surface or the potential of mean force of the protein, $F(\{\Delta\mathbf{R}\})$, defined by Eq. (7.1). In order to clarify the physical meaning of $\mathbf{f}_\alpha$, we define the free energy of the entire system in terms of the statistical mechanics by the configurational integral as follows,

$$F_\mathrm{s} = -kT\log\int\cdots\int \exp\left[-\frac{1}{k_\mathrm{B}T}U(\{\mathbf{R}\},\{\mathbf{r}\})\right]\prod_{\alpha=1}^{N_\mathrm{u}} d\mathbf{R}_\alpha \prod_{i=1}^{N}\prod_{a=1}^{n} d\mathbf{r}_i^a \tag{7.9}$$

where $\{\mathbf{R}\}$ and $\{\mathbf{r}\}$ denote sets of coordinates of atoms in protein and water respectively. $U(\{\mathbf{R}\},\{\mathbf{r}\})$ is the interaction energy among the atoms in the system, $\mathbf{R}_\alpha$ denotes the coordinate of the αth atom in protein, and $\mathbf{r}_i^a$ labels the ath atom in the ith water molecule. Taking the derivative of F_s with respect to the position of an atom in protein, we get the following expression,

$$-\left(\frac{\partial F_s}{\partial \mathbf{R}_\alpha}\right) = \frac{\int\cdots\int(-\partial U/\partial \mathbf{R}_\alpha)\exp[-U(\{\mathbf{R}\},\{\mathbf{r}\})/k_B T]\prod_{\alpha=1}^{N_u} d\mathbf{R}_\alpha \prod_{i=1}^{N}\prod_{a=1}^{n} d\mathbf{r}_i^a}{\int\cdots\int\exp[-U(\{\mathbf{R}\},\{\mathbf{r}\})/k_B T]\prod_{\alpha=1}^{N_u} d\mathbf{R}_\alpha \prod_{i=1}^{N}\prod_{a=1}^{n} d\mathbf{r}_i^a} \tag{7.10}$$

The expression indicates that $\mathbf{f}_\alpha$ is a mean force acting on the atom α due to all other atoms in protein and in solvent, averaged over the statistical ensemble.

Structural Fluctuation Induced by Pressure: The thermodynamic work due to pressure is one of the perturbation that induces the structural change of protein. With the consideration in mind, we decouple the derivative into two steps as follows:

$$\left(\frac{\partial F_s}{\partial \mathbf{R}_\alpha}\right) = \left(\frac{\partial F_s}{\partial V}\right)_T \left(\frac{\partial V}{\partial \mathbf{R}_\alpha}\right)_{P,T} \tag{7.11}$$

where V denotes the volume of the system. The first factor in the right-hand side is related to the pressure in terms of the standard thermodynamics.

$$-\left(\frac{\partial F_s}{\partial V}\right)_T = P, \tag{7.12}$$

while the second factor is identified as

$$\left(\frac{\partial V}{\partial \mathbf{R}_\alpha}\right)_{P,T} = \left(\frac{\partial \Delta\overline{V}}{\partial \mathbf{R}_\alpha}\right)_{P,T}, \tag{7.13}$$

where $\Delta\overline{V}$ is the partial molar volume of a protein in solution. It is resulted from an a priori consideration that the change in the volume due to the structural change is only concerned with the partial molar volume. Combined Eqs. (7.6) and (7.11)–(7.13) together give rise to the following equation:

$$\langle \Delta\mathbf{R}_\alpha \rangle = (k_B T)^{-1} \sum_\beta \langle \Delta\mathbf{R}_\alpha \Delta\mathbf{R}_\beta \rangle_0 \cdot \left(\frac{\partial \Delta\overline{V}}{\partial \mathbf{R}_\beta}\right)_{P,T} P \tag{7.14}$$

The equation provides a physical explanation concerning how the hydrostatic pressure applied to the system induces the structural change, or fluctuation in protein.

According to the equation, the structural response to pressure is realized through the successive response functions. The first one is the first derivative of the partial molar volume with respect to the atomic coordinates of protein. The second is the variance–covariance matrix of the fluctuation in the unperturbed system. When pressure is applied, the deformation first occurs so that the partial molar volume decreases. Then, the deformation is propagated to the entire protein through the variance–covariance matrix.

Three important points are worthwhile to be noted with respect to (7.14). The first is that the both response functions are *intrinsic* properties of a protein solution, meaning that the properties are determined just by the interactions among atoms in protein and water. The pressure plays a role of magnifying the intrinsic properties.

The second is that the direction of the structural change, $D \rightleftarrows N$, is determined essentially by the sign of $\left(\partial \Delta \overline{V} / \partial \mathbf{R}_\beta\right)_{P,T}$ which is the change in the partial molar volume due to the spatial fluctuation of atom β in protein.

The third point to be noted is that all the quantities included in the expression, $\left\langle \Delta \mathbf{R}_\alpha \Delta \mathbf{R}_\beta \right\rangle$, $\Delta \overline{V}$ and $\left(\partial \Delta \overline{V} / \partial \mathbf{R}_\beta\right)_{P,T}$ can be calculated by means of RISM/3D-RISM theory.

Structural Fluctuation Induced by Heat: The deformation due to heat is the most well-studied property of protein. In this case, the change in the entropy is the perturbation that induces the structural change of protein, and it is the energy that contains the entropy as the independent variable. The energy is expressed by the first law of thermodynamics.

$$\mathrm{d}E = T\mathrm{d}S - p\mathrm{d}V \tag{7.15}$$

Then, the perturbation produced on the protein structure by heat is defined as

$$\mathbf{f}_\alpha = -\frac{\mathrm{d}E}{\mathrm{d}\mathbf{R}_\alpha} = -\left(\frac{\mathrm{d}E}{\mathrm{d}S}\right)_V \left(\frac{\mathrm{d}S}{\mathrm{d}\mathbf{R}_\alpha}\right) = -T\left(\frac{\mathrm{d}\Delta\overline{S}}{\mathrm{d}\mathbf{R}_\alpha}\right) \tag{7.16}$$

where $\overline{S}$ denotes the partial molar entropy of protein, which is also a functional of protein structure. Combining Eq. (7.6) with Eq. (7.16), one finds

$$\langle \Delta \mathbf{R}_\alpha \rangle = -(k_\mathrm{B} T_0)^{-1} \sum_\beta \left\langle \Delta \mathbf{R}_\alpha \Delta \mathbf{R}_\beta \right\rangle_0 \left(\frac{\mathrm{d}\overline{S}}{\mathrm{d}\mathbf{R}_\beta}\right) T \tag{7.17}$$

where T_0 denotes the temperature concerned with the reference system.

The expression tells us how the structural fluctuation of protein due to heat or temperature is coupled with the change in the partial molar entropy of protein. As was clarified in Chap. 4, the change in the partial molar entropy consists of two major contributions: one from the *structural entropy*, the other from solvent, that is,

$$S_{\text{conf}} = \frac{1}{2} k_{\text{B}} \log\left(\frac{(2\pi)^{3N}}{A}\right) + \frac{3N}{2} k_{\text{B}} \tag{7.18}$$

$$s_{\text{solv}}(\mathbf{r}, \mathbf{r}') = \frac{1}{2} k_{\text{B}} \log\left(\frac{(2\pi)^{n}}{B}\right) + \frac{3n}{2} k_{\text{B}} \tag{7.19}$$

where A is the determinant of the variance–covariance matrix defined by Eq. (7.4), and B is essentially the determinant of the inverse pair-correlation-function-matrix of solvent (see Chap. 4 for detailed definition).

The partial molar entropy is defined as the sum of the two contributions as

$$\overline{S} \equiv S_{\text{conf}} + S_{\text{solv}} \tag{7.20}$$

where S_{solv} is defined by

$$S_{\text{solv}} \equiv \int_V \int_V s_{\text{solv}}(\mathbf{r}, \mathbf{r}') \mathrm{d}\mathbf{r} \mathrm{d}\mathbf{r}' \tag{7.21}$$

Equations (7.17)–(7.21) suggest us the physics of the heat/cold denaturation. The direction of structural change due to rising/lowering temperature will be determined by the sign of $(\mathrm{d}\overline{S}/\mathrm{d}\mathbf{R}_\beta)$ in (7.20). The partial molar entropy $\overline{S}$ consists of two contributions, the conformational entropy S_{conf} and the solvent entropy S_{solv}.

It is not so difficult to guess just by intuition that S_{conf} will increase with increasing the structural fluctuation $\Delta\mathbf{R}_\beta$. On the other hand, it is not so obvious whether S_{solv} increases or decreases upon the fluctuation, because it is conceivable that there are two competing contributions to the quantity upon increasing structural fluctuation.

The native state of protein has many water molecules inside its cavities such as active sites of an enzyme. If those water molecules are released from the cavity upon denaturation, then they will acquire entropy. Therefore, such process of water release will make the contribution to entropy *positive*. It may be the physics of so-called "heat denaturation."

On the other hand, the denaturation of protein may cause the hydrophilic residues, such as glutamic acid and lysine, exposed to solvent. Such processes will make the entropy change of solvent upon denaturation negative, since the strong hydration due to the hydrophilic residues will reduce the degrees of freedom of water molecules. The physics may explain the phenomena so-called "cold denaturation".

Structural Fluctuation Induced by Denaturant: Structural change is induced by adding the denaturant such as urea into the system. It is obviously the change in the chemical potential that drives the structural change due to denaturant. The free energy change induced by the change in the concentration of solution may be expressed by,

$$\mathrm{d}F = \sum_i (N_i \mathrm{d}\mu_i + \mu_i \mathrm{d}N_i) \tag{7.22}$$

where N_i and μ_i are the number of molecules (or concentration) and the chemical potential of the species i included in the solution.

Now we just do a thought experiment in which concentration of a species regarded as the "denaturant" is changed, keeping the concentration of other species, including water and protein, constant. Let label the species by subscript d.

Then, the free energy changes due to adding denaturant is expressed as,

$$\mathrm{d}F = \mu_d \mathrm{d}N_d + N_d \mathrm{d}\mu_d. \tag{7.23}$$

We define the perturbation due to the structural fluctuation by Eq. (7.8) or

$$\mathbf{f}_\alpha \equiv -\left(\frac{\partial F}{\partial \mathbf{R}_\alpha}\right)_{T,V}. \tag{7.24}$$

Substituting Eq. (7.23) into Eq. (7.24), one finds

$$\mathbf{f}_\alpha = -\left(\frac{\partial \mu_d}{\partial \mathbf{R}_\alpha}\right)_{T,V}, \tag{7.25}$$

because the concentration of denaturant (N_d) does not change by the structural change ($\mathrm{d}\mathbf{R}_\alpha$).

Here, it should be noted that μ_d is an implicit functional of the protein structure $\{\mathbf{R}\}$, because the direct and solvent-induced interactions between protein and denaturant depends sensitively on the structure. Inserting Eq. (7.25) into Eq. (7.7), one gets

$$\langle \Delta \mathbf{R}_\alpha \rangle = -(k_\mathrm{B}T)^{-1} \sum_\beta \langle \Delta \mathbf{R}_\alpha \Delta \mathbf{R}_\beta \rangle_0 \left(\frac{\partial \mu_d}{\partial \mathbf{R}_\beta}\right) N_d \tag{7.26}$$

According to the equation, the perturbation of the denaturant on the structure of protein is created first through a response of the structure to the change in the chemical potential of denaturant. Then, the structural response is propagated over the entire molecule through the variance–covariance matrix.

The expression is telling us that the sign of $(\partial \mu_d / \partial \mathbf{R}_\beta)$ determines the direction of the structural change whether the protein unfolds by adding the denaturant or not. Here, μ_d is the chemical potential of denaturant species d in the aqueous solution, and it is a function of the protein structure, because the chemical potential depends on the protein structure in the vicinity of protein.

Extending the Theory to the Protein Folding: It is likely that the variance–covariance matrix in Eq. (7.26) changes significantly due to a perturbation, even though the distribution still remains Gaussian.

One such example is the thermal denaturation processes described by Eq. (7.17). The variance–covariance matrix for the denatured state is apparently much greater than that of the native state, since the conformational fluctuation for the denatured state should be much greater than the native state due to the increased entropy. The linear response theory, Eq. (7.6), in its original form may not be applied to those processes. However, we may be able to extend the theory to such a case by means of the *analytical continuation* described above.

We divide the perturbation and the response into several steps for which the linear response theory is valid. We refer to such a process as a "quasi-linear" process. The process is defined by Eq. (7.7), that is,

$$\langle \Delta \mathbf{R}_\alpha \rangle_{j+1} = \left(k_\mathrm{B} T_j\right)^{-1} \sum_\beta \left\langle \Delta \mathbf{R}_\alpha \Delta \mathbf{R}_\beta \right\rangle_j \cdot \mathbf{f}_\beta$$

where j denotes each quasi-linear step, starting from zero. In the equation, the variance–covariance matrix at the $(j+1)$th step should be calculated based on the protein coordinates in the jth step.

With a proper care made for "scheduling" of increasing the order parameter, we may be able to get successive curves similar to Fig. 7.2, which have different values for the curvature, or the variance–covariance matrix.

7.2 Structural Transition Induced by a Local Chemical/ Mechanical Perturbation in Biomolecules

There are physicochemical processes commonly observed in a biomolecule, in which a local perturbation applied to the structure induces a global conformational change of the molecule. The conformational change so induced gives large impact on functions which biomolecules play in a variety of life phenomena (Terazima 2016).

The first example of such processes is found in the biosynthesis carried out by a bacterium, in which a conformational change of the *retinal* moiety bound at an amino-acid residue of bacteriorhodopsin, from *cis* to *trans*, triggers a structural change in the host molecule, which in turn induces successive chemical reactions including the electron transfer at the photo-active center (Voet and Voet 1995).

The next example of such structural changes, caused by a chemical/mechanical perturbation, is found in an amino-acid substitution, or *mutation* of a protein. The amino-acid substitution triggers a change in the global conformation. The change in global conformation in turn causes loss or acquisition of activity that the protein plays. It is a common strategy for a microorganism to use the mutation to survive against a drug. Therefore, it is crucial for medical motivation to find the structural change of target proteins, induced by the amino-acid substitution (Phanich et al. 2016).

The third case is the intramolecular charge-transfer reaction in protein, an example of which is the electron transfer taking place in the photo-active center in plant cells.

In the preceding section, we have presented a linear response theory to predict a global change of biomolecules, such as protein folding, induced by the change in the thermodynamic variables such as temperature, pressure, and concentration of denaturant molecules.

In the following, we formulate a theoretical framework to predict the conformational reorganization of a biomolecule caused by a *chemical* or/and *mechanical* perturbation induced at a *moiety*, or a localized portion, of the molecule. It is assumed that the conformational change thus introduced does not change the inherent activity of the biomolecule (Hirata 2023a).

Structural Transition Induced by a Local Conformational Change in Biomolecules: The theory to describe the structural change of a biomolecule induced by a *local* structural change is formulated based on the linear response theory described in the preceding section, or Eq. (7.7),

$$\langle \Delta \mathbf{R}_\alpha \rangle_1 = (k_B T)^{-1} \sum_\beta \langle \Delta \mathbf{R}_\alpha \Delta \mathbf{R}_\beta \rangle_0 \cdot \mathbf{f}_\beta \tag{7.27}$$

The local structural change induces changes in the potential energy U of the biomolecule, which includes that of solute and solvent. For formulating the linear response theory, the potential energy is decomposed into two contributions; the reference part and the perturbation,

$$U(\{\mathbf{R}\}) = U_r(\{\mathbf{R}\}_r) + \Delta U \tag{7.28}$$

where $U_r(\{\mathbf{R}\}_r)$ represents the interaction energy of the reference part, and ΔU is the perturbation. The perturbation part consists of three contributions as

$$\Delta U = \Delta U_m(\{\mathbf{R}\}_m) + \Delta U_{rm}(\{\mathbf{R}\}_r, \{\mathbf{R}\}_m) + \Delta U_{solv}, \tag{7.29}$$

where ΔU_m is the change in potential energy among atoms in the moiety, ΔU_{rm} is that between the moiety and the reference part of protein, and ΔU_{solv} represents the change of the solute–solvent interaction, induced by the perturbation.

It is worthwhile to note that ΔU_{solv} may play important role especially in the case of an intramolecular charge-transfer reaction. It is because such a reaction is associated with a significant reorganization of solvent around the biomolecule, and the energy of solvent reorganization is an important factor to determine the rate of electron transfer reaction (Marcus and Sutin 1985; Chong et al. 1995).

The perturbative force can be expressed as,

$$\mathbf{f}_\beta = -\frac{\partial \Delta U}{\partial \mathbf{R}_\beta} = -\left(\frac{\partial \Delta U_{rm}(\{\mathbf{R}\}_r, \{\mathbf{R}\}_m)}{\partial \mathbf{R}_\beta} + \frac{\partial \Delta U_{solv}}{\partial \mathbf{R}_\beta} \right) \tag{7.30}$$

where $\mathbf{R}_\beta$ represents the coordinate of an atom in the reference system. The derivative of ΔU_m is disappeared, because it is not concerned with the coordinate of the atom β in the reference part of protein.

Combining Eqs. (7.27) and (7.30), one finds

Fig. 7.3 Schematic picture of the reaction: the quadratic curves depict the free energy surface of reactant and product, and $\Delta U(\{\mathbf{R}\})$ represents the perturbation defined by Eq. (7.29)

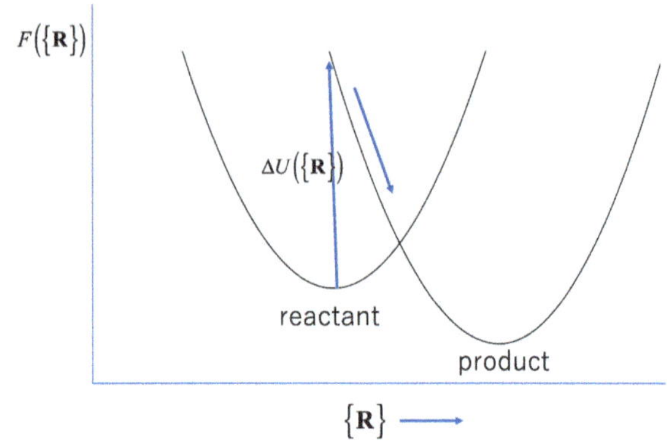

$$\langle \Delta \mathbf{R}_\alpha \rangle_1 = -\frac{1}{k_B T} \sum_\beta \left\langle \Delta \mathbf{R}_\alpha \Delta \mathbf{R}_\beta \right\rangle_0 \cdot \frac{\partial \Delta U}{\partial \mathbf{R}_\beta}, \tag{7.31}$$

where $\left\langle \Delta \mathbf{R}_\alpha \Delta \mathbf{R}_\beta \right\rangle_0$ is the variance–covariance matrix of the reference system, or, the protein without the moiety, and the force $\partial \Delta U / \partial \mathbf{R}_\beta$ is defined by (7.30).

Physical meaning of the expression, Eq. (7.31), is as follows. *The force exerted by atoms in the moiety causes the displacement in atom β of protein. The displacement propagates through the variance–covariance matrix $\left\langle \Delta \mathbf{R}_\alpha \Delta \mathbf{R}_\beta \right\rangle_0$ to induce a global structural change of the molecule, $\langle \Delta \mathbf{R}_\alpha \rangle_1$.*

By the perturbation introduced at a portion of protein, some conformation change of the biomolecule will be induced. But, the structural change should not be so large, by definition, as for the biomolecule to keep its inherent activity.

Such a local perturbation introduced at a biomolecule may cause changes of the free energy surface illustrated by the following picture (Fig. 7.3). The figure illustrates that the equilibrium structure shifts from reactant to product due to the perturbation introduced at a small portion of a biomolecule.

A question may be raised with respect to the applicability of Eq. (7.31) to a process of mechanical and/or chemical perturbation. What if a large perturbation is applied, the response to that may not necessarily be explained by the linear response theory? In such a case, the native conformation of the biomolecule may be lost. There are many such cases found in life phenomena, in which native structure of a biomolecule is deformed due to a chemical or mechanical perturbation. In the present section, we are not concerned with such cases, because the biomolecule will lose its function anyway in cases. Only proteins that have some biological function are qualified to survive in nature. In other words, our target in the present section is a native protein in which the structural fluctuation is *linear* against a perturbation.

There may be another question concerning the computational procedure to apply the theory to conformational change of protein. Since a protein suffers a quite large perturbation as amino-acid substitution, the linear-response scheme of the theory may fail.

However, it is largely an issue of the computational protocol. The problem may be resolved by using the *analytical continuation* explained in the previous section.

In the procedure, the entire mechanical/chemical perturbation is separated into N steps, so that each step can be treated within the linear response theory, or Eq. (7.31). Then, the overall structural change after N-steps may be represented by,

$$\langle \Delta \mathbf{R}_\alpha \rangle = \frac{1}{k_B T} \sum_{j=1}^{N} \sum_{\beta} \langle \Delta \mathbf{R}_\alpha \Delta \mathbf{R}_\beta \rangle_j \cdot \left(-\frac{\partial \Delta U_{\mathrm{rm}}^{(j)}(\{\mathbf{R}\}_{\mathrm{r}}, \{\mathbf{R}\}_{\mathrm{m}})}{\partial \mathbf{R}_\beta} \right). \tag{7.32}$$

A careful choice of N, or *scheduling*, may be required for an actual implementation of the theory to a biophysical process (Hirata 2023a).

Examples of Possible Application of the Theory: Examples of possible applications of the theory proposed in the present section to bio-physical and/or -chemical process are discussed in the following.

One of the applications is concerned with the *drug discovery*. It is well documented that a crucial step for designing a drug compounds, either by a wet chemistry or in silico, is not only to identify a target protein, but also to find its molecular structure. Current state of the art in the business is to determine the structure by means of the X-ray crystallography, or the two-dimensional NMR, both of which are quite laborious as well as time consuming. For examples, in case of the X-ray crystallography, one has to crystalize the molecule first to be able to get the diffraction pattern, from which one may extract the conformation. If there are multiple candidate-proteins for a drug target, the process becomes formidable some time. If the mutation of the micro-organism is very quick, such efforts become entirely hopeless, as was demonstrated unambiguously by the latest pandemic of COVID-19. The theory explained here can be applied to such a problem to find the structure of a target protein, which is derived from the native protein by an amino-acid substitution. We still need to identify which amino-acid is substituted by which. However, such an identification would be neither very difficult nor time-consuming by means of current state of the art, for examples, by the NMR spectroscopy.

Another example of the application is the mutagenesis to improve the activity of an enzyme, or to add a new activity to the biocatalyst (Voet and Voet 1995). In that case, too, it is important to confirm that the native conformation of the enzyme is intact after substituting an amino-acid by a new one. If the biomolecule is denatured by the mutation, it will lose or reduce the activity as a biocatalyst. Since the amino-acid substitution has been currently made in trial and error basis by means of a wet chemistry, the process becomes quite laborious as well as time consuming. Therefore, it will be desired to predict the structure of the biomolecule after each trial of an amino-acid substitution employing the method proposed in the present section.

Based on the Kim–Hirata theory for the structural fluctuation of a biomolecule in aqueous solution, a theoretical method was presented to predict the conformation of the biomolecule, a moiety of which is modified chemically or/and mechanically by such a process as photo-excitation or amino-acid substitution. The method is expected to be applied to the medical as well as biomimetic processes in industry, for the purpose of accelerating the process to find the conformation of biomolecules after chemical or/and mechanical modifications. It is the key of success for the application to employ the RISM/3D-RISM method to calculate the free energy surface of the protein, including *solvation free energy*, and its second derivative with respect to the atomic coordinate of the biomolecule.

The benefit of using the RISM/3D-RISM is not limited to the application described in the present paper. Using the 3D-RISM theory in a series of chemical processes may accelerate an industrial innovation, for examples, the discovery of a new drug. Let us mention two such examples in which the use of RISM/3D-RISM theory may accelerate industrial processes.

One of such processes is concerned with the discovery of new medicines. The process of the drug discovery consists of many steps; finding a protein to be targeted, screening the candidate compounds for the target protein, synthesizing the molecules, analyzing solubility of the molecule, clinical examination, and so forth. The computational approach may play an important role in the compound screening. It is to find a chemical compound among many candidate molecules, which exhibits the highest affinity to the targeted protein. Malfunctioning the protein may be fatal for the host micro-organism.

The RISM/3D-RISM method has been applied by several research groups, including authors', targeting many proteins to predict the binding affinity of candidate compounds (Imai et al. 2009, 2011; Sugita and Hirata 2016; Hasegawa et al. 2017; Hayashino et al. 2018; Sugita et al. 2020, 2021). However, the applications have been largely limited so far to the cases in which the conformation of the target protein has been already provided in the protein data bank (PDB). If the conformation is unknown, the method has been entirely "out of scope". The method described in the present section may be valid for such cases, because the method can predict the conformation of protein created by mutations. The screening of drug compounds may be dramatically accelerated, because the wet chemistry to determine the structure of mutant protein is replaced by the in-silico process.

The other process, to which the present theory may be applied, is the rate of chemical reaction in biomolecular solutions. An example of such reactions is the photochemical reaction, in which the reaction induced at a chromophore, such as a cis–trans isomerization of retinal, triggers a successive conformational change in the entire protein. The initial photochemical reaction is so fast that the conformational change of entire protein becomes the rate-determining step.

In the following section, an Arrhenius-type theory of the reaction rate will be introduced, in which the activation barrier is defined by the crossing point of two parabolas representing the free energy surfaces of the reactant and product along the reaction coordinate (Hirata 2023b).

7.3 Rate of Chemical Reactions of Biomolecules in Solution: Generalized Langevin Mode Analysis (GLMA)

In the preceding sections, we have developed a theory that can be applied to conformational change, or *isomerization reaction*, of a biomolecule, induced by a thermodynamic perturbation such as pressure, temperature, and denaturant. We have also formulated a theory to explore reactions induced by a local chemical and/or mechanical perturbation in biomolecules (Hirata and Akasaka 2015; Hirata 2023a).

It has been clarified that the thermodynamic perturbation produces a new free-energy surface which is also quadratic with respect to the displacement of atoms of the biomolecule. Then, we have two free energy surfaces before and after the *reaction*, both are quadratic, with different minima as well as curvatures. If one draws the free energy surfaces along the structural fluctuation as reaction coordinates, the two surfaces will intersect to form an activation barrier. It is a natural extension of the concept to develop a theory for the rate of chemical reaction based on the barrier-crossing model.

In the following, let us review briefly a little history of the activation-barrier-crossing-model of chemical reactions.

A Brief History of the Barrier-Crossing Model of Chemical Reactions: The rate of chemical reaction in a system of constant volume can be expressed by the following equation,

$$k_{\text{rate}} = \nu \exp\left(- \Delta F^{(\dagger)}/k_B T\right) \tag{7.33}$$

where $\Delta F^{(\dagger)}$ is the activation free-energy in general, and ν is the pre-factor that takes account for other effects such as solvent viscosity. The theory has been applied to a wide variety of experimental studies for understanding the essential physics involved in the reaction rate.

It was well regarded among experimentalists that the logarithm of the chemical-reaction-rate plotted against inverse temperature falls on a straight line with negative slope,

$$\log k_{\text{rate}} \propto -(1/T).$$

The behavior now known as *Arrhenius plot* was first brought to the thermodynamic consideration by van't Hoff (Laidler 1963). Then, Arrhenius forwarded the idea further to complete the formulation now popularly stated as "Arrhenius theory" (Arrhenius 1884). In the following, we briefly review the Arrhenius theory in a slightly different way.

Let's think of a chemical equilibrium,

$$C^{(\mathrm{r})} \underset{k_-}{\overset{k_+}{\rightleftarrows}} C^{(\mathrm{p})} \tag{7.34}$$

where $C^{(\mathrm{r})}$ and $C^{(\mathrm{p})}$ represent the concentration of the reactant and product, respectively, and k_+ and k_- denote the rate of reaction in the respective direction. The equilibrium constant K_{eq} of the reaction is defined by

$$K_{\mathrm{eq}} \equiv \frac{k_+}{k_-}. \tag{7.35}$$

The equilibrium constant is related to the difference in the free energy between the two states as,

$$K_{\mathrm{eq}} = \exp\left[-\frac{\Delta F_{\mathrm{eq}}}{k_{\mathrm{B}}T}\right], \tag{7.36}$$

$$\Delta F_{\mathrm{eq}} = F_{\mathrm{eq}}^{(\mathrm{p})} - F_{\mathrm{eq}}^{(\mathrm{r})}, \tag{7.37}$$

where $F_{\mathrm{eq}}^{(\mathrm{r})}$ and $F_{\mathrm{eq}}^{(\mathrm{p})}$ represent the free energy of reactant and product states, respectively. Taking the logarithm, one finds

$$\ln K_{\mathrm{eq}} = \ln k_+ - \ln k_- = -\left(F_{\mathrm{eq}}^{(\mathrm{p})} - F_{\mathrm{eq}}^{(\mathrm{r})}\right)/k_{\mathrm{B}}T. \tag{7.38}$$

Now, we *assume* that the reactant and the activated states are in equilibrium with the activated state, respectively, that is,

$$C^{(\mathrm{r})} \underset{k_{\dagger\to\mathrm{r}}}{\overset{k_{\mathrm{r}\to\dagger}}{\rightleftharpoons}} C^{(\dagger)}, \quad C^{(\dagger)} \underset{k_{\mathrm{p}\to\dagger}}{\overset{k_{\dagger\to\mathrm{p}}}{\rightleftharpoons}} C^{(\mathrm{p})}. \tag{7.39}$$

Then, we find

$$\begin{aligned} \ln K_{\mathrm{eq}}^{\mathrm{r}\dagger} &= \ln k_{\mathrm{r}\to\dagger} - \ln k_{\dagger\to\mathrm{r}} \\ \ln K_{\mathrm{eq}}^{\mathrm{p}\dagger} &= \ln k_{\mathrm{p}\to\dagger} - \ln k_{\dagger\to\mathrm{p}} \end{aligned} \tag{7.40}$$

The activation free energy is defined by the crossing point of two parabolas, reactant and product, which is the height of free energies measured from the respective equilibrium-states. That consideration leads the following relation for the activation free energies.

$$\begin{aligned} \Delta F_{\mathrm{r}}^{\dagger} &= F^{(\dagger)} - F_{\mathrm{eq}}^{(\mathrm{r})} \quad \text{or} \quad F_{\mathrm{eq}}^{(\mathrm{r})} = F^{(\dagger)} - \Delta F_{\mathrm{r}}^{\dagger} \\ \Delta F_{\mathrm{p}}^{\dagger} &= F^{(\dagger)} - F_{\mathrm{eq}}^{(\mathrm{p})} \quad \text{or} \quad F_{\mathrm{eq}}^{(\mathrm{p})} = F^{(\dagger)} - \Delta F_{\mathrm{p}}^{\dagger} \end{aligned} \tag{7.41}$$

From Eqs. (7.38)–(7.41), one finds,

$$
\begin{aligned}
\ln k_{+} - \ln k_{-} &= -\left(F_{\text{eq}}^{(\text{p})} - F_{\text{eq}}^{(\text{r})}\right)/k_{\text{B}}T \\
&= \left(F^{(\dagger)} - \Delta F_{\text{r}}^{\dagger}\right)/k_{\text{B}}T - \left(F^{(\dagger)} - \Delta F_{\text{p}}^{\dagger}\right)/k_{\text{B}}T \\
&= -\Delta F_{\text{r}}^{\dagger}/k_{\text{B}}T + \Delta F_{\text{p}}^{\dagger}/k_{\text{B}}T
\end{aligned} \tag{7.42}
$$

The forward and backward reactions are independent each other. Therefore, one finds,

$$
\log k_{+} = -\frac{\Delta F_{\text{r}}^{\dagger}}{k_{\text{B}}T} + \text{const}, \quad \log k_{-} = -\frac{\Delta F_{\text{p}}^{\dagger}}{k_{\text{B}}T} + \text{const}. \tag{7.43}
$$

The theoretical formulation of the barrier-crossing model of chemical reaction in *gas phase* has been made by Eyring, referred to as "absolute rate theory" (Glasstone et al. 1941). Two assumptions have been made for the physical model of the theory. Firstly, the *energy* surface of the molecules involved in the reaction is *harmonic*. Secondly, the reactant and activated states of the molecules are in *chemical equilibrium*. Based on the two assumptions, Eyring has derived the following expression for diatomic molecules in gas phase,

$$
k_{\text{rate}} = \frac{kT}{h}\frac{Q^{(\dagger)}}{Q_A Q_B}\exp\left(-\Delta E^{(\dagger)}/k_{\text{B}}T\right) \tag{7.44}
$$

where Q_A and Q_B denote the vibrational partition-functions of the diatomic molecules, A and B, in reactant, while $Q^{(\dagger)}$ is that of activated states, respectively. In the expression, $\Delta E^{(\dagger)}$ is essentially an electronic energy that can be evaluated thoroughly from the quantum chemistry in principle. The expression, however, cannot be applied to chemical reactions in biomolecular systems, which inevitably involve the effect of solvent on the reactions.

Kramers have formulated the rate of chemical reaction in *solution* based on the Smoluchowski equation for the barrier-crossing model. He has taken account for the viscosity of solvent (Kramers 1940).

$$
k_{\text{rate}} = \nu(\eta)\exp\left(-\Delta F^{(\dagger)}/k_{\text{B}}T\right) \tag{7.45}
$$

where $\Delta F^{(\dagger)}$ is the activation energy and $\nu(\eta)$ is the pre-factor that takes the viscosity η of solvent into consideration. The factor plays an important role especially in a reaction in which solvent viscosity makes significant contribution. The expression, however, may not be applied to a chemical reaction of a biomolecule in water, since it does not involve the contribution from the *solvation free-energy* to the *activation free-energy*.

For chemical reactions in biomolecular solutions, it is the crucial step to find the contribution from solvent fluctuation to activation barrier, since $\Delta F^{(\dagger)}$ is the *free energy* of the crossing point of the two parabolas that represent the structural as well as solvent-density fluctuation of reactant and product states of the biomolecule. The idea to involve solvent-density fluctuation was embodied by Marcus in the electron-transfer reaction in solutions (Marcus and Sutin 1985). In the theory, the barrier is defined as the crossing point of two *free-energy* surfaces which are attributed to the states before and after the electron transfer. It is assumed that the both free-energy surfaces are *quadratic* or *harmonic* with respect to the fluctuation of electronic *charges* on the reactant and product molecules or ions. The idea has its basis on the expression proposed by Born for the free energy of ions in polar liquids, which is a quadratic function of ionic charges (Born 1920). The polarization of solvent around ions and the free energy are fluctuating around an average value. The free energy is expressed by the quadratic function of the charges, and the polarization or the response function is expressed as a function of dielectric constant of solvent. When the polarization of solvent is not in equilibrium with the electronic charge, the free energy climbs up along the quadratic slope in a reactant state, and cross over the activation barrier to transfer on the free energy surface of the product state.

A molecular theory of the electron-transfer reaction was proposed by Chong et al. in 1995, in which polarization of water around ions are treated based on the statistical mechanics of liquids, or the XRISM theory (Chong et al. 1995). In the theory, the non-equilibrium free-energy surface of an ion, which represents the solvent fluctuation, was produced by changing the ionic charge as an order parameter. The method was combined with the RISM-SCF theory, an electronic structure theory of a solvated molecule, to analyze the rate of electron-transfer reaction of small molecules in water (Tenno et al. 1993; Sato et al. 2003, 2004; Yoshida et al. 2008). The structural fluctuation, or *isomerization*, of molecules was not considered in those analyses, because the structural fluctuation of such a small molecule is so fast compared to the solvent fluctuation associated with the electron-transfer reaction. However, such a treatment may not be valid in a biomolecule in solution, because the structural fluctuation or isomerization reaction is slow, and it is the rate determining step of the reaction.

In the present section, we consider the rate of chemical reaction referred to as "isomerization" of a biomolecule in aqueous solutions. By "isomerization" reaction, we mean a conformational change of a molecule, which is not associated with a recombination of chemical bonds. The protein folding is the most outstanding example of the isomerization reaction in biomolecules. The electron-transfer reaction in protein at the photo-active-center is considered to be another example of such a reaction, because the reaction does not cause a recombination of chemical-bonds, but is associated with a significant change in the solvated structure of the protein (Sasaki 2016).

Structural Fluctuation in Biomolecular Reactions: A life is maintained by a variety of chemical reactions in solution, such as photo synthesis, enzymatic reactions, and so forth (Voet and Voet 1995). The time constant of those reactions ranges from nanoseconds to minutes or hours in some cases. Why the time constant of the reactions is so widely spread is because the reaction is always associated with a reorganization of the molecular structure, or *isomerization*, of proteins in which the reaction is taking place. A reaction will not be completed until the structure of biomolecules and its environment, or *solvent polarization*, are equilibrated *thermodynamically* to the product state of the reaction. Here, "thermodynamically" means that the product state should be at a minimum of the *free energy* surface including the *solvation* free energy, not just the *potential energy* of a *naked* biomolecule. Although the time constant of the chemical reactions is widely spread depending on the reaction, they are well separated in many cases of interest. In the following, let us sketch few examples of such cases.

So-called *protein folding* is an *isomerization* reaction in which the many dihedral-angles around chemical bonds constructing the backbone structure should change harmoniously from gauche to trans or vice versa. It is a global change of the structure. On the other hand, the global change of structure is associated with many *local isomerization reactions* with a variety of time constants (Anfinsen 1973; Akasaka 2006; Hirata et al. 2018).

Let us see the separation of the time constant in the enzymatic reactions. An enzymatic reaction consists essentially two *chemical processes* as been clarified by Michaelis and Menten: binding or unbinding of a substrate at a cavity of protein (enzyme), and a *chemical reaction* or recombination of atoms in substrate molecules inside the cavity (Michaelis and Menten 1913). In general, time constants of those processes are well separated. In fact, it is the basic assumption in the Michaelis–Menten theory that the reaction between bound and unbound states, or the *molecular recognition process*, is so quick that they are already in *equilibrium*. However, the molecular recognition process itself has some time-constant which may be short, but it is of current interest in life science (Sugita et al. 2021).

A group of proteins called "molecular channels" has an interesting mechanism, referred to as "gating" which plays a role of *valve*, controlling the binding/unbinding of small ions and molecules to be transported to/from the channel pore. The gating is nothing but a local conformational change, or *isomerization*, of the biomolecule (Hille 2001).

Another example of the time separation in a chemical process in protein is seen in an *energy-transfer* or/and *charge-transfer* reaction, induced by photo excitations. A typical example of such reactions is the twisted intramolecular charge-transfer (TICT) reaction, in which the charge transfer is conjugated with a local conformational change, or "twisting", of the molecule. In such a case, the rate of local conformational change largely determines the reaction rate (Sasaki et al. 2016).

As such, a conformational change of a biomolecule is a composite of many *isomerization reactions*, each of which has different rate-constants. Therefore, it is the crucial issue to decompose the entire reaction into independent reactions, each of which has different time constant.

Decoupling of Fluctuational Modes and Reaction Coordinates: Recently, a theory is formulated based on the GLMA theory to calculate the rate constant of chemical reactions taking place in biomolecules in solution (Hirata 2023b).

In the theory, a chemical reaction in solution is regarded as a "barrier crossing" problem, similar to the Marcus theory of the electron transfer reaction. The activation barrier is defined as a crossing point of two free-energy surfaces, each of which is attributed to the reactant and product of the reaction. The most essential premise of the theory is that the both free-energy surfaces, reactant and product, are "quadratic" or "harmonic". The premise concerning the free energy surface is based on the generalized Langevin equation for structural fluctuation of protein derived in Chap. 4, according to which the force acting on each atom in the equation is linear to their displacement from the equilibrium state (Kim and Hirata 2013). The quadratic surface is a composite of many harmonic functions with different modes or frequencies, which are *hierarchically* ordered. The height of the activation barrier will be also ordered hierarchically: higher the frequency, higher the barrier. So, the reaction rate should be analyzed according to the mode of fluctuation. It is the fundamental concept on which the new theory is based.

Mode Decoupling of the Free Energy-Surface: The free energy surface in the Cartesian-coordinate space described in the previous section is a composite of many surfaces with different curvature, or *mode*, which has a hierarchical construction. It will be rational to think that the rate of a chemical reaction taking place in a macromolecule such as protein is highly dependent on the *mode* of structural fluctuation.

As an example, let's think of the folding reaction of a protein molecule. Imagine a prototypical folding-reaction, or a two-states transition from denatured to native states. The free energy surface of such a *reaction* is represented by two curves with wells along a reaction coordinate, each of which is expressed by a quadratic function. The crossing point of the two curves is identified as the *transition state* of the reaction. Apparently, the height of the activation barrier primarily determines the rate of reaction. It should be noted, however, that the overall reaction-rate of folding is a composite of many *isomerization-reactions*, for examples, gauche-trans isomerization around dihedral angles, having different timescales hierarchically ordered depending on their modes. Such *chemical reactions* with different modes may also be of interest for exploring the rate of reactions taking place in protein, such as in an enzyme.

It is the advantage of the Generalized Langevin mode analysis (GLMA) to decompose the mode, and to project the free energy surface onto a coordinate axis of a single mode. Such decoupling of modes can be made by diagonalizing the Hessian matrix $\{K_{\alpha\beta}\}$ defined by

$$K_{\alpha\beta} = \frac{\partial^2 F}{\partial \mathbf{R}_\alpha \partial \mathbf{R}_\beta}. \tag{7.46}$$

As has been shown in Eq. (4.54) in Chap. 4, the Hessian matrix is an inverse of the variance–covariance matrix,

$$K_{\alpha\beta} \equiv k_B T \left(\mathbf{L}^{-1}\right)_{\alpha\beta}, \tag{7.47}$$

where $\mathbf{L}$ is the variance–covariance matrix defined by $\mathbf{L} \equiv \langle \Delta\mathbf{R}\Delta\mathbf{R} \rangle$.

The diagonalization of the matrix is carried out by a standard linear-algebra in the multi-dimensional space. First of all, we must find the origin of the coordinate space, at which the free energy of the molecule becomes minimum in the configuration space spanned by the $3N-6$ atomic coordinates. (The six degrees of freedom removed from the analysis are those related to the translational and rotational motions of the biomolecule as a whole.) The origin can be found by translating the origin of the configuration space to the center of the free energy surface. Alternatively, the origin can be found by minimizing the free energy surface, where the two origins should coincide by definition (see Chap. 5).

Then, the axis is *rotated* to get the new axis which diagonalizes the Hessian matrix. The free energy projected onto the new coordinate-space, or the *normal-mode* coordinates space, can be expressed in the standard form as

$$F(\{Q\}) = \sum_{m}^{3N-6} K_m\left(Q_m - Q_{m,\mathrm{eq}}\right)^2 \tag{7.48}$$

where Q_m denotes the normal-mode coordinates of the mth mode, $Q_{m,\mathrm{eq}}$ is the coordinate at the origin, or the equilibrium structure, and K_m represents the Hessian or the "force constant" associated with the mth mode.

GLMA for the Rate of Chemical Reactions in Solution: Equation (7.48) is applied to formulate the rate of chemical reaction in terms of the activation-barrier crossing. First of all, we have to define the mode which is *relevant to* the chemical reaction. We put the following hypothesis.

The mode whose center of coordinate and/or the frequency changes most before and after the reaction is the one relevant to the chemical reaction, and choose the coordinate as the reaction coordinate.

For examples, the coordinates corresponding to the least eigen value is likely to be the reaction coordinate in case of the *protein folding*.

Let us denote the two free energy surfaces for reactant and product, projected on to the reaction coordinate by,

$$F^{(\mathrm{r})}(Q) = K^{(\mathrm{r})}\left(Q - Q_{\mathrm{eq}}^{(\mathrm{r})}\right)^2 + F_{\mathrm{eq}}^{(\mathrm{r})} \tag{7.49}$$

$$F^{(\mathrm{p})}(Q) = K^{(\mathrm{p})}\left(Q - Q_{\mathrm{eq}}^{(\mathrm{p})}\right)^2 + F_{\mathrm{eq}}^{(\mathrm{p})} \tag{7.50}$$

where the suffices "r" and "p" specify the quantities concerning reactant and product, and $F_{\mathrm{eq}}^{(\mathrm{r})}$ and $F_{\mathrm{eq}}^{(\mathrm{p})}$ in the equations represent the free energy of the equilibrium structure of the reactant and product systems (see Fig. 7.4).

The barrier is defined as a crossing point of the two parabolas expressed by Eqs. (7.49) and (7.50). It is a trivial mathematical problem to find the crossing point of two simple parabolas like Eqs. (7.49) and (7.50). Let's denote the crossing point of the coordinate of the mth normal-mode by $Q^{\dagger}$. In the special case in which the

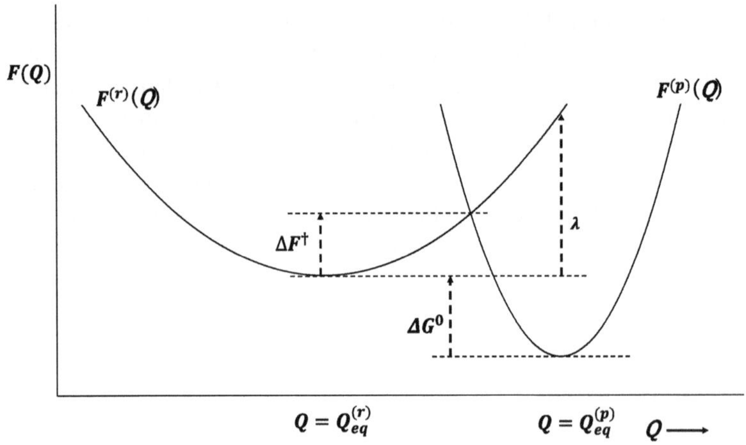

Fig. 7.4 Illustration of the free energy surface along the reaction coordinate

curvatures of the two parabollas are the same,

$$\Delta F^{(\dagger)} \equiv F^{(r)}\left(Q^{\dagger}\right) - F^{(r)}\left(Q_{eq}^{(r)}\right) \tag{7.51}$$

So-called *energy gap* ΔG^0 and the solvent reorganization energy, which are relevant to the chemical reaction, can be also defined by the following equations,

$$\Delta G^0 \equiv F_{eq}^{(p)} - F_{eq}^{(r)} \tag{7.52}$$

$$\lambda = F^{(r)}\left(Q_{eq}^{(p)}\right) - F^{(r)}\left(Q_{eq}^{(r)}\right) \tag{7.53}$$

where λ is referred to as the *reorganization energy* in the Marcus theory of the electron-transfer reaction (Marcus and Sutin 1985; Chong 1995). It is noted that $\Delta G^0 < 0$ and $\lambda > 0$ in ordinary chemical reactions.

Finally, the rate of reaction is given by

$$k_{rate} = \nu(\zeta_m)\exp\left[-\frac{\Delta F^{(\dagger)}}{k_B T}\right] \tag{7.54}$$

In the special case in which the curvatures of the two parabollas are the same,

$$\Delta F^{(\dagger)} = \frac{\left(\lambda + \Delta G^0\right)^2}{4\lambda} \tag{7.55}$$

which is the familiar expression given by Marcus for the electron transfer reaction (Marcus and Sutin 1985; Chong et al. 1995).

The pre-factor $\nu(\zeta_m)$ included in (7.54) represents the factor to determine the reaction rate other than the height of the activation barrier. In the case of *isomerization* reaction with which this chapter is concerned, the factor is essentially governed by the friction due to solvent. It should be noted that the friction ζ_m is dependent on mode as was clarified in Chap. 5. (See Chap. 5 for the friction exerted on a temporal structure of protein.)

An Illustrative Example of Possible Applications of the Theory: In the following, some idea will be provided for understanding how the theory can be applied for characterizing chemical reactions in biomolecules in solution. The process of protein folding has been roughly categorized into the three groups: (a) two-states transition, (b) barrier less transition, and (c) multi-states transition.

The two-state transition is the most popular model in which the unfolded state of protein should cross over a single free-energy barrier in order for folding into the native conformation. The free energy surface along the reaction coordinate in that case is illustrated in Fig. 7.5a, which is more or less similar to the case drawn in Fig. 7.4. That is, the free-energy surface corresponding to the unfolded state has the greater variance reflecting the larger entropy, while that corresponding to the folded state has deeper well, implying the greater attractive interactions, may be, due to hydrogen-bonds between backbone atoms.

The second case is the so-called "barrier-less transition" from denatured to native conformations. In that case, the equilibrium structures between the two states should be very different, thereby the minima of the two free-energy parabola should be separated far away along the reaction coordinate Q. It is also imagined that the curvature of the free-energy-parabola in the denatured conformation is so small that activation barrier virtually disappears. Illustrated in Fig. 7.5b is such a situation. In the figure, the parabola $F^{(u)}(Q)$ corresponding to the denatured state is so shallow that $\Delta F^{(\dagger)}/k_B T$ virtually disappears.

Then, why does the curvature of the free energy surface become so small? The question can be answered by looking back Eqs. (7.46) and (7.47), which define the Hessian of the free energy surface. As shown in Eq. (7.47), the Hessian matrix is an inverse of the variance–covariance matrix. Therefore, small Hessian means large variance. Then, what is thermodynamic-quantity related to the variance–covariance matrix? The answer is the *structural entropy* of protein. As shown in Eq. (4.90) in Chap. 4, greater the variance–covariance matrix, larger the structural entropy. Thus, we reach the following conclusion. It is the condition for the barrier-less folding of protein that unfolded state of protein has large structural entropy.

The conclusion with respect to the barrier-less folding has been reached already by Chong and Ham. They have made an analysis of the MD trajectory of a protein, produced by ANTON (Chong and Ham 2011). The authors have calculated the structural entropy of the protein along the trajectory based on the RISM/3D-RISM theory, and have found that the transition is *barrier-less* in accord with the *energy-landscape model* proposed by Wolynes and his coworkers (Bryngelson et al. 1995).

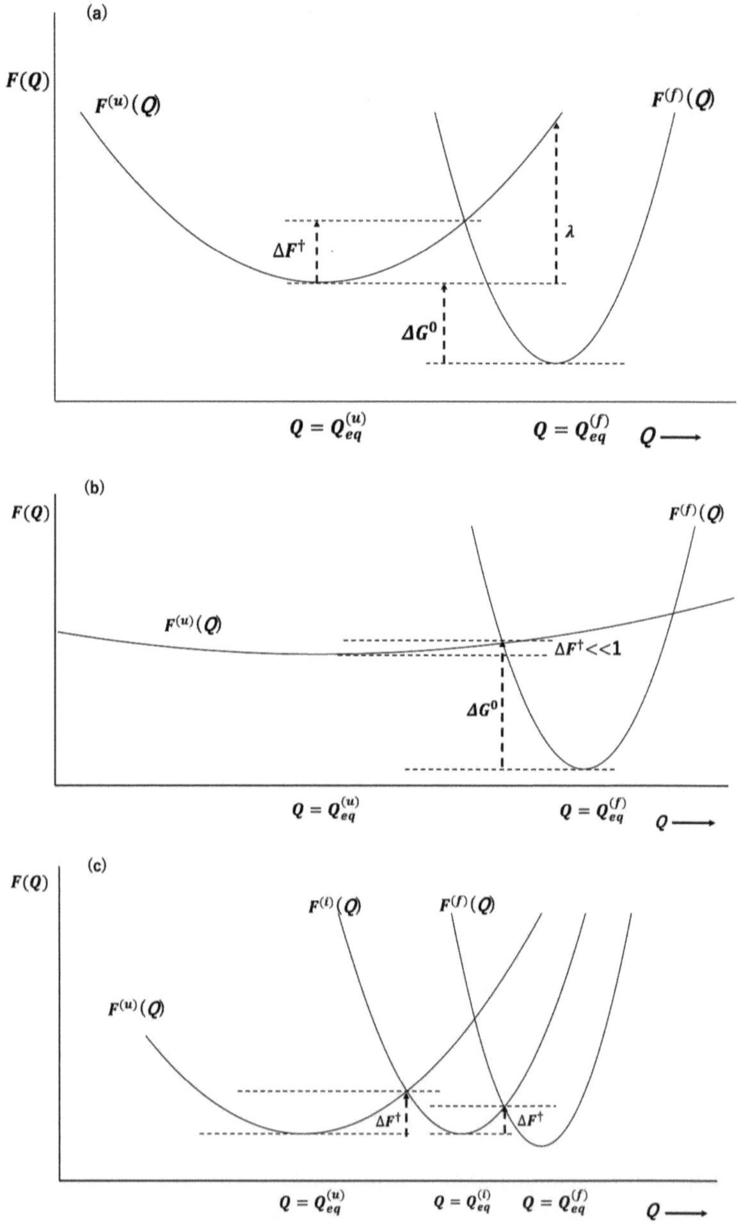

Fig. 7.5 Illustration of the three cases of the protein folding-reaction: superscripts "u", "f" and "i" denote "unfolded", "folded", and intermediate states, respectively. **a** The two-states model, **b** the barrier-less folding, **c** the folding through an intermediate state

They have concluded that the large structural-entropy of the denatured state of protein is the origin of the barrier-less transition.

The third type of the protein folding involves an intermediate state along the folding pathway. In one of such cases, an intermediate conformation referred to as "molten-globule" structure is involved. In the case, most of the secondary structures are already formed as in the native conformation. However, *packing* of those structure into the native conformation is not completed (Arai and Kuwajima 2000). The observation suggests that the difference in the three-dimensional conformations between the reactant and the intermediate state may not be so different. Therefore, the curvatures of the parabola between the two may not be so different as well. Nevertheless, the reaction should pass through an activation barrier, which may not be so high, but not negligible. Such a situation is illustrated in Fig. 7.5c.

Perspective: In the present section, the latest development made by the author and his coworkers in chemical reactions in biomolecular system was reviewed. It is formulated on the basis of the generalized Langevin theory combined with the RISM/3D-RISM theory of molecular liquids. The theory analyzes the reaction rate based on rather traditional view, or the transition state theory, developed by the great scientists in physical chemistry such as Arrhenius, Eyring, Kramers, and Marcus. According to the theory, the reaction rate is determined essentially by the activation barrier which is formed by the two free-energy surfaces, each of which describes the conformational fluctuation of the respective state of molecules, reactant or product. The structural fluctuation of biomolecules in solution is characterized by the free energy surfaces, especially, by its second derivatives with respect to the atomic coordinates or *Hessian*. It is suggested, as a possible application of the theory, to investigate the folding mechanism of protein, for examples, whether it is a *barrier crossing* process between the two states or a *barrier-less* process.

In the present book, our attention was just focused on so-called "isomerization reaction" or "tautomerization" in which the reaction proceeds along an adiabatic surface. Protein folding is the most outstanding example of such reactions. The reactions are concerned neither with the recombination of the chemical bond nor with the *non-adiabatic* transition between the two surfaces. Enzymatic reactions are the best example of such reactions not included in the scope of the theory. Electron-transfer reactions at photo-active center of plant is another example in which the non-adiabatic transition plays a crucial role determining the rate of reaction. In those examples, (classical) structural fluctuation and change in electronic structure are highly conjugated.

In order to account for the physics of a chemical reaction in biomolecules, the three elements in theoretical chemistry are required in addition to the theory presented in the present book: (1) a theory to evaluate the electronic structure of a molecule in *aqueous solutions*; (2) a computational method to calculate the electronic structure of a *macromolecule* such as protein; and (3) a theory to describe the *non-adiabatic transition* between two adiabatic surfaces. If such theoretical elements are put together, all the physicochemical processes in life phenomena will become under control.

A theory to explore the electronic-structure of a molecule in aqueous solutions has been already proposed three decades ago (Tenno et al. 1993; Sato et al. 1996). The theory referred to as "RISM-SCF" has been successfully applied to many problems related to chemical reactions, including the electron-transfer reactions in water (Yoshida et al. 2008). Recently, the method has been extended to macromolecules such as DNA (Yoshida and Hirata 2006; Yoshida et al. 2011).

It has been a non-trivial problem to calculate the electronic structure of a macromolecule such as protein, due mainly to the computational time. However, the problem has become less serious lately in two reasons: the development of high-performance computer, and the development of models or/and methodologies in the quantum chemistry such as the ONIOM developed by Morokuma and his coworkers, and the fragment MO (FMO) developed by Kitaura and his coworkers (Chung et al. 2015; Kitaura et al. 1999).

A considerable progress has been made during the past few decades by H. Nakamura and his coworkers concerning the non-adiabatic transition, which has further developed the idea proposed by Landau and Zener (Landau 1932; Zener 1932; Nakamura 2019).

It will be certainly a great challenge for computational science to put those theories and methodologies all together to calculate the rate of chemical reactions taking place in biomolecules. Nevertheless, it will be a feasible task considering the latest development in the information technology both in hardware and software.

7.4 Ion Transport Through Ion Channels

In our body, a class of protein referred to as "molecular transporter" is playing vital roles to transport molecules as well as *information* among cells. Ion channels and aquaporins (water channel) are the transporters which are most ubiquitously existing in our body. In the present section, we formulate a theory to describe the conductivity of a molecular transporter from the view point of the barrier-crossing model, taking ion channels as an example.

As been shown in Fig. 1.1b in the introduction section of the book, the free energy surface along the pathway of ions inside an ion-channel consists of alternating series of tops and bottoms. When an ion travels through the channel, it should climb up every top of the free-energy hill, and the height of the top will be the crucial factor to determine the rate of permeation, or the conductivity. The problem may be formulated from the view point of the barrier-crossing model.

The rate or conductivity of an ion to pass through the channel may be defined by,

$$\Lambda = L/\tau \tag{7.56}$$

where L denotes the distance between the entrance and exit of the channel, and τ the time spent by an ion to pass through the channel. L and τ may be expressed by the following relations,

$$L = \sum_i l_i, \quad \tau = \sum_i \tau_i, \tag{7.57}$$

where l_i is the distance between the two basins in the channel pore, and τ_i denotes the time elapsed during the barrier crossing. Then, the rate to pass over the ith activation barrier can be expressed by,

$$\lambda_i = l_i/\tau_i. \tag{7.58}$$

Since l_i and L is given experimentally, it is the elapsed time τ that should be estimated theoretically. From Eqs. (7.57) and (7.58), one finds

$$\tau = \sum_i \tau_i = \sum_i (l_i/\lambda_i) \tag{7.59}$$

Therefore, it is the rate λ_i for passing over the ith barrier along the channel that should be evaluated theoretically.

Here, we adopt the barrier-crossing model to evaluate the rate λ_i. Then, the rate may be expressed by the Arrhenius equation as,

$$\lambda_i = \lambda_0 \exp\left(-\frac{\Delta F_i^{\dagger}}{k_B T}\right) \tag{7.60}$$

where $\Delta F_i^{\dagger}$ is the activation barrier between the two successive bottoms. The pre-factor λ_0 signifies the conductivity of the ion in aqueous solutions without the activation barrier. The conductivity can be evaluated by means of the generalized Langevin theory combined with the RISM theory. Such a theory has been proposed by Chong and Hirata (1999). Therefore, it is the activation free energy $\Delta F_i^{\dagger}$ that should be evaluated theoretically in order to complete the theory.

The activation free energy may be defined by the following equation.

$$\Delta F_i^{\dagger} = F_i^{\mathrm{top}} - F_i^{\mathrm{bottom}}, \tag{7.61}$$

where F_i^{bottom} and F_i^{top} are the bottom and top of the ith hill on the free energy surface, projected along the central axis of the channel pore.

The free energy surface along the channel pore is evaluated from the density distribution function $g(\mathbf{r})$ of ions based on the following equation,

$$F(\mathbf{r}) = -k_B T \log g(\mathbf{r}). \tag{7.62}$$

As an example, the distribution functions and the free energy profiles of water and ions in the potassium channel, calculated by means of the RISM/3D-RISM theory, are depicted, respectively, in Fig. 1.1a, b. As one can see, the free energy profile

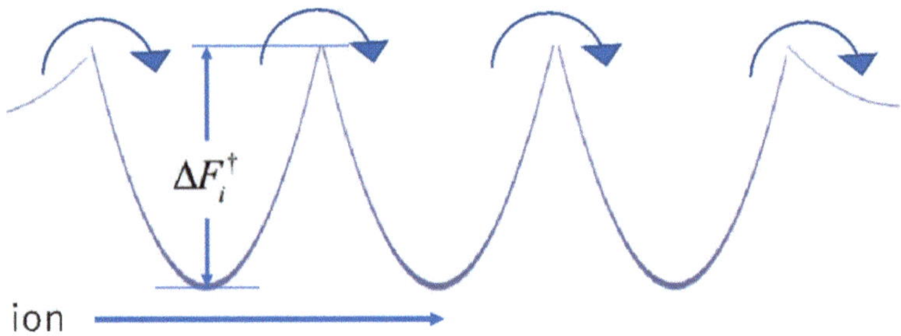

Fig. 7.6 Image of ion conduction through an ion channel: an ion passes through a channel by crossing over a series of activation barriers on the free energy surface inside the channel pore

consists of a series of hills, and F_i^{bottom} and F_i^{top} are defined as the bottom and top of the ith free-energy hill.

Image of the barrier crossing model of ion conduction through ion channels is illustrated in Fig. 7.6.

References

Akasaka K (2006) Probing conformational fluctuation of proteins by pressure perturbation. Chem Rev 106:1814–1835

Anfinsen CB (1973) Principle that govern the folding of protein chain. Science 181:223–230

Arai M, Kuwajima K (2000) Role of the molten globule state in protein folding. Adv Protein Chem 53:209–282

Arrhenius S (1884) Recherches sur la conductivité galvanique des électrolytes. Doctoral dissertation, Royal Publishing House, P.A. Norstedt & Söner, Stockholm

Born VM (1920) Volumen und Hydrationswarme der Ionen. Z Phys 1:45–48

Bryngelson JD, Onuchic JN, Socci ND, Wolynes PG (1995) Funnels, pathways, and the energy landscape of protein folding: a synthesis. Proteins 21:167–195

Chong S-H, Ham S (2011) Configurational entropy of protein: a combined approach based on molecular simulation and integral-equation theory of liquids. Chem Phys Lett 504(46):225–229

Chong S-H, Hirata F (1999) Dynamics of ions in liquid water: an interaction-site-model description. J Chem Phys 111(8):3654–3667

Chong S-H, Miura S, Basu G, Hirata F (1995) Molecular theory for the non-equilibrium free energy profile in electron transfer reaction. J Phys Chem 99:10526

Chung LW, Sameera WMC, Ramozzi R, Page AJ, Hatanaka M, Petrova GP, Harris TV, Li X, Ke Z, Liu F, Li H, Morokuma K (2015) The ONIOM method and its applications. Chem Rev 115:5678–5796

Friedman HL (1985) A course in statistical mechanics. Prentice-Hall, New Jersey

Glasstone K, Laidler KJ, Eyring H (1941) The theory of rate processes. McGraw-Hill, New York

Hasegawa T, Sugita M, Kikuchi T, Hirata F (2017). A systematic analysis of the binding affinity between the Pim-1-kinase and its inhibitors based on the MM/3D-RISM/KH method. J Chem Info Model 57:2789–2798

Hayashino Y, Sugita M, Arima H, Irie T, Kikuchi T, Hirata F (2018) Predicting the binding mode of 2-hydroxypropyl-beta-cyclodextrin to cholesterol by means of the MD simulation and the 3D-RISM-KH theory. J Phys Chem B 122:5716–5725

Hille B (2001) Ion channels of excitable membranes, 3rd edn. Sinauer

Hirata F (2023a) Structural transition induced by a local chemical/mechanical perturbation in biomolecules. Condens Matter Phys 26(43803):1–8. https://doi.org/10.5488/CMP.26.43803
Hirata F (2023b) A theory of chemical reactions in biomolecules in solution: generalized Langevin mode analysis (GLMA). J Chem Phys 158. https://doi.org/10.1063/5.0143849.Str
Hirata F, Akasaka K (2015) The structural fluctuation of proteins induced by thermodynamic perturbation. J Chem Phys 044110
Hirata F, Sugita M, Yoshida M, Akasaka K (2018) Perspective: structural fluctuation of protein and Anfinsen's thermodynamic hypothesis. J Chem Phys 148:020901
Ikeguchi M, Ueno J, Sato M, Kidera A (2005) Protein structural change upon ligand binding: linear response theory. Phys Rev Lett 94:078102
Imai T, Oda K, Kovalenko A, Hirata F, Kidera A (2009) Ligand mapping on protein surfaces by the 3D-RISM theory: toward computational fragment-based drug design. J Am Chem Soc 131:12430–12440
Imai T, Miyashita N, Sugita Y, Kovalenko A, Hirata F, Kidera A (2011) Functionality mapping on internal surfaces of multidrug transporter AcrB based on molecular theory of solvation: implications for drug efflux pathway. J Phys Chem B 115:8288–8295
Kim B, Hirata F (2013) Structural fluctuation of protein in water around its native state: a new statistical mechanics formulation. J Chem Phys 138:054108–054111
Kitaura E, Ikeo E, Asada T, Uebayashi M (1999) Fragment molecular orbital method: an approximate computational method for large molecules. Chem Phys Lett 313:701–706
Kramers HA (1940) Brownian motion in a field of force and the diffusion model of chemical reactions. Physica 7:284–304
Kubo R, Toda M (1991) Statistical physics II: nonequilibrium statistical physics. Springer
Laidler KJ (1963) Reaction kinetics: volume I—homogeneous gas reaction. Pergamon Press, London
Landau DL (1932) Phys Zts Sov 2:46
Marcus RA, Sutin N (1985) Electron transfer in chemistry and biology. Biochim Biophys 811:265–322
Michaelis L, Menten M (1913) Die kinetik der invertinwirkung. Biochem Z 49:333–369
Nakamura H (2019) Introduction to non-adiabatic dynamics. World Scientific
Phanich J, Rungrotmongkol T, Sindhikara D, Phongphanphanee S, Yoshida N, Hirata F, Kungwan N, Hannonbua S (2016) A 3D-RISM/RISM study of the oseltamivir binding efficiency with the wild-type and resistance-associated mutant forms of the viral influenza B neuraminidase. Protein Sci 25:147–158
Sasaki S, Drummen GPC, Konishi G (2016) Recent advances in twisted intramolecular charge transfer (TICT) fluorescence and related phenomena in materials chemistry. J Mater Chem C 4:2731–2743
Sato H, Hirata F, Kato S (1996) Analytical energy gradient for the reference interaction site model multiconfigurational self-consistent-field method: application to 1,2-difluoroethylene in aqueous solution. J Chem Phys 105:1546
Sato H, Kobori Y, Tero-Kubota S, Hirata F (2003) Theoretical study on electronic and solvent reorganization associated with a charging process of organic compounds: I. Molecular and atomic level description of solvent reorganization. J Chem Phys 119:2753–2760
Sato H, Kobori Y, Tero-Kubota S, Hirata F (2004) Theoretical study on electronic and solvent reorganization associated with a charging process of organic compounds: 2. A new decomposition procedure into electrostatic and nonelectrostatic responses. J Phys Chem B 108:11709–11715
Sugita M, Hirata F (2016) Predicting the binding free energy of the inclusion process of 2-hydroxypropyl-b-cyclodextrin and small molecules by means of the MM/3D-RISM method. J Phys Condens Matter 28:384002–384013
Sugita M, Hamano M, Kasahara K, Kikuchi T, Hirata F (2020) New protocol for predicting the ligand-binding and mode based on the 3D-RISM/KH theory. J Chem Theor Comput 16:2864–2876

Sugita M, Kuwano I, Higashi T, Motoyama K, Arima H, Hirata F (2021) Computational screening of a functional cyclodextrin derivative for suppressing a side effect of doxorubicin. J Phys Chem B 125:2308–2316
Sugita M, Onishi I, Irisa M, Yoshida Y, Hirata F (2021) Molecular recognition and self-organization in life phenomena studied by a statistical mechanics of molecular liquids, the RISM/3D-RISM theory. Molecule 26:271–300
Tenno S, Hirata F, Kato S (1993) A hybrid approach for the solvent effect on the electronic structure of a solute based on the RISM and Hartree–Fock equations. Chem Phys Lett 214:391–396
Terazima M (2016) Molecular science of fluctuations toward biological functions. Springer, Japan
Voet D, Voet JG (1995) Biochemistry. Wiley, New York
Yoshida N, Hirata F (2006) A new method to determine electrostatic potential around a macromolecule in solution from molecular wave functions. J Comput Chem 27(4):453–462
Yoshida N, Ishida T, Hirata F (2008) Theoretical study of temperature and solvent dependence of the free energy surface of the intramolecular electron transfer based on the RISM-SCF theory; application to 1,3-dinitrobenzene radical anion in acetonitrile and methanol. J Phys Chem 112:433–440
Yoshida N, Kiyota Y, Hirata F (2011) The electronic-structure theory of a large-molecular system in solution: application to the intercalation of proflavine with solvated DNA. J Mol Liq 159:83–92
Zener C (1932) Non-adiabatic crossing of energy levels. Proc R Soc A A137:696–702

Correction to: Transitions Among Equilibrium States Driven by Fluctuations (Perspective)

Correction to:
Chapter 7 in: F. Hirata, *Molecular Theory of Fluctuation in Life Phenomena*,
https://doi.org/10.1007/978-981-96-3364-7_7

The original version of the book was inadvertently published with an error in Page 138 of Chapter 7 which has been corrected. The book and the chapter have been updated with the changes.

The updated version of this chapter can be found at
https://doi.org/10.1007/978-981-96-3364-7_7

F. Hirata, *Molecular Theory of Fluctuation in Life Phenomena*,
https://doi.org/10.1007/978-981-96-3364-7_8

MIX
Papier aus verantwortungsvollen Quellen
Paper from responsible sources
FSC® C105338

If you have any concerns about our products,
you can contact us on
ProductSafety@springernature.com

In case Publisher is established outside the EU,
the EU authorized representative is:
Springer Nature Customer Service Center GmbH
Europaplatz 3, 69115 Heidelberg, Germany

Printed by Libri Plureos GmbH
in Hamburg, Germany